W. E. Russey, C. Bliefert, C. Villain

Text and Graphics
in the Electronic Age

VCH

© VCH Verlagsgesellschaft mbH, D-69451 Weinheim (Federal Republic of Germany), 1995

Distribution:

VCH, P. O. Box 10 1161, D-69451 Weinheim (Federal Republic of Germany)

Switzerland: VCH, P. O. Box, CH-4020 Basel (Switzerland)

United Kingdom and Ireland: VCH (UK) Ltd., 8 Wellington Court,
    Cambridge CB1 1HZ (England)

USA and Canada: VCH, 220 East 23rd Street, New York, NY 10010–4606 (USA)

Japan: VCH, Eikow Building, 10-9 Hongo 1-chome, Bunkyo-ku, Tokyo 113 (Japan)

ISBN 3-527-28519-9

William E. Russey, Claus Bliefert
Christophe Villain

Text and Graphics
in the Electronic Age

Desktop Publishing for Scientists

Weinheim · New York
Basel · Cambridge · Tokyo

Dr. William E. Russey
Professor of Chemistry
Juniata College
Huntingdon, PA 16652
USA

Christophe Villain
3, Rue de Pins Moisy
F-92190 Meudon
France

Prof. Dr. Claus Bliefert
Labor für Umweltchemie
Fachbereich Chemieingenieurwesen
Fachhochschule Münster
Stegerwaldstraße 39
D-48565 Steinfurt
Federal Republic of Germany

Library of Congress Card No. applied for

British Library Cataloguing-in-Publication Data:
A catalogue record for this book is available from the British Library

Deutsche Bibliothek Cataloguing-in-Publication Data
Russey, William E.:
Text and graphics in the electronic age : desktop publishing for scientists /
William E. Russey ; Claus Bliefert ; Christophe Villain. –
Weinheim ; New York ; Basel ; Cambridge ; Tokyo : VCH, 1995
ISBN 3-527-28519-9
NE: Bliefert, Claus:; Villain, Christophe

© VCH Verlagsgesellschaft mbH, D-69451 Weinheim (Federal Republic of Germany), 1995

Printed on acid-free and low-chlorine paper

Composition: BCU Software, D-48620 Schöppingen
Printing: betz-druck GmbH, D-64291 Darmstadt
Bookbinding: Großbuchbinderei J. Schäffer, D-67269 Grünstadt
Printed in the Federal Republic of Germany

To our friend Hans F. Ebel

Preface

Modern techniques of electronic word processing now make it possible for authors to prepare their own camera-ready copy even for a major book project—consistent with the highest standards of professional quality and relatively modest outlays for software and hardware. It is therefore not surprising that the roles of publishing houses and printers have changed dramatically in recent years, and that further progress in "desktop publishing" promises even more sweeping change in the years ahead.

This raises the important questions: What contribution can reasonably be expected from the average scientist or engineer in industry or academia with respect to the preparation of manuscripts destined for publication, taking into account both the time factor and material investment? What in fact are the limitations associated with these new avenues to publication? What skills and information are required? What is absolutely necessary in the way of hardware and software? How do the results of "desktop publishing" really compare to the products of professional publishers? These and related questions constitute the central issues in our book.

The best computer in the world, supported by outstanding publishing software, is incapable of providing superior results in the absence of solid technical knowledge and practiced skills on the part of the user. In order to take full advantage of the potential inherent in a computer and the accompanying software, a "desktop publisher" must become familiar not only with strengths and weaknesses of the technical system at his or her disposal, but also with the basic principles of traditional typesetting. It is for this reason that we have chosen to divide our book into three major segments.

In Part I, *Basic Principles*, we begin by discussing the fundamental nature of electronic manuscripts and what it is we mean by "desktop publishing" (Chapter 1). This is followed by a consideration of some of the ground rules of typography and layout (Chapter 2), as well as the special problems posed by scientific and mathematical text (Chapter 3). Our approach is quite general, and completely independent of any decision with respect to a particular hardware or software platform. At the same time, however, the *information* presented is crucial, and its mastery should be regarded as a fundamental prerequisite for the layman aspiring to prepare camera-ready copy that will be acceptable in professional circles.

Part II provides insight into selected types of *hardware* available for home-processing of text and graphics. Here we discuss not only the technical considerations but also the most effective ways of utilizing computers (Chapter 4), storage devices (Chapter 5), printers (Chapters 6 and 7), and scanners (Chapter 8). In this section and throughout the book it is not our intent to dwell on particular advantages or disadvantages of specific products, but rather to provide a general introduction to the most important features of the various components of a personal computer system optimized for desktop-publishing applications. Many of the examples we discuss are associated with the Apple MACINTOSH hardware platform, the system we ourselves use and one that has proven over the years to be especially well-suited to our needs, but we have also made a conscientious attempt to draw meaningful parallels to the popular WINDOWS environment for IBM-compatible PC systems.

Part III is devoted to *software* useful in a desktop-publishing context. We have tried especially hard to illustrate the wide range of choices open to the amateur publisher, devoting roughly equal attention to text and graphics. The section opens with a consideration of the role of word processors (Chapter 9), followed by an extensive discussion of various types of graphics software (Chapter 10). This leads naturally to a relatively thorough examination of programs for combining text and graphics (Chapter 11). The book concludes with an overview of "miscellaneous software" likely to be of interest to many "scientific desktop publishers".

A few words are perhaps in order regarding the origins of this book. Many years of stimulating collaboration with respect to scientific communication culminated in our first public presentation, Ebel/Bliefert/Russey: *The Art of Scientific Writing* (1987; reprinted: 1990), sponsored by the publisher responsible for the present volume as well, VCH. Intense personal involvement with the special challenges posed by camera-ready copy in the natural sciences led to the subsequent publication (1989, in German) of Bliefert/Villain: *Text und Grafik*, which was released two years later in French translation by the Paris publishing house Technique & Documentation-Lavoisier under the title *La PAO scientifique*.

Like the present book, which contains a great deal of new material, the two earlier volumes constituted a serious attempt to expose a wider audience to the exciting possibilities open to the scientific community as a result of recent progress in the techniques of "desktop publishing" (what the French call "Publication Assistée par Ordinateur", PAO).

* * *

We take this opportunity to express our very special debt of gratitude to an important close friend: Dr. Hans F. Ebel, for many years Senior Editor at VCH in Weinheim. Frequent long and engaging conversations with Hans over the course of many years served as an important catalyst in the maturation of the thoughts and insights collected in these pages. More important, however, it was through the constant, gentle prodding of Hans that this book finally came to see the light of day, a book that clearly reflects much of his own immense store of wisdom, experience, and foresight.

Huntingdon, Schöppingen,   William E. Russey
and Paris, October 1994   Claus Bliefert
                                                  Christophe Villain

Contents

Part III: Software

9   *Word Processing*   231

9.1   Introduction   231
9.2   Minimum Requirements for Convenient Text Preparation   234
9.3   Special Considerations with Respect to Scientific Text   243
9.4   MICROSOFT WORD   246
9.4.1   Functions   246
9.4.2   Graphics and Equations   252
9.5   Other Word-Processing Programs   254
9.6   File Formats for Text   256
9.7   Mathematical and Physical Formulas and Equations   258
9.8   General Guidelines for Writing with a Word Processor   260

10   *Page-Layout Programs*   263

10.1   Introduction   263
10.2   PAGEMAKER   266
10.3   QUARKXPRESS   273
10.4   Other Page-Layout Programs   275
10.5   Typesetting Software   280

11   *Graphics Programs*   283

11.1   Introduction   283
11.2   Bitmap-Oriented Graphics Programs   285
11.3   Object-Oriented Graphics Programs   289
11.4   PostScript-Oriented Graphics Programs   294
11.5   3D-Graphics and CAD Programs   298
11.6   Chemical Formulas   299

12   *Miscellaneous Software*   305

12.1   Introduction   305
12.2   Database Programs   306
12.2.1 General Observations   306
12.2.2 Flat-File Database Software   308
12.2.3 Relational Databases   311

Appendixes

Part I

Basic Principles

1 Electronic Manuscripts and "Desktop Publishing"

1.1 The Electronic Manuscript

1.1.1 Introduction

Electronic manuscripts—computer-generated documents preserved in electromagnetic form—represent a relatively recent phenomenon, but their role in society is already large, and it is continuing to expand at a dramatic rate. Publishers of books and periodicals, indeed businesses of all types, are increasingly subject to the sweep of the electronic document revolution, and the impact has been no less profound with respect to the individual writer. The electronic trend is perhaps most apparent with respect to text, but it also encompasses graphic images of every conceivable type. Enthusiastic converts to the new technology are constantly challenged to take additional steps in the direction of professional document presentation by acquiring the very latest tools, offering ever greater potential to reward creative efforts.

In today's world the long-favored *typewritten* document (or *typescript*) is on the verge of becoming an anachronism. Especially in the case of a document destined ultimately to become part of a book or a journal, a manuscript in paper form is likely to be regarded as little more than a temporary repository for changes waiting to be recorded in the electronic counterpart. Perhaps the only significant exception to this sweeping generalization is the meticulously prepared printed manuscript designed to serve as *camera-ready copy*, a final processing stage immediately prior to photomechanical reproduction and subsequent wider distribution (cf. Sec. 1.1.6).

An *electronic* manuscript[1] (sometimes known as a *machine-readable* manuscript or a *compuscript*) differs in one fundamental and crucial way from the traditional paper manuscript or typescript: all the characters and symbols of which it is composed actually exist only within the fragile and narrow confines of an electromagnetic storage medium; extracting them for examination or modification always requires the use of a sophisticated data-processing device. The "data processor" most familiar to authors in this context is the *personal computer*, and the most important storage medium is usually the *diskette* ("floppy disk").

This particular aspect of the overall electronic revolution has been responsible for profound changes in the publishing industry.[2] Many of the developments initially met with considerable skepticism, but publishers quickly came to appreciate one particular consequence of the trend: diskettes are far more convenient to send, receive, and process than bulky paper manuscripts. Perhaps more important, however, electronic documents offer the promise of eliminating much of the time-consuming, costly, and error-prone drudgery of conventional typesetting.

An interesting parallel development is worth at least passing mention, because it too is likely soon to have an impact on scientist–authors: the fact that certain types of information are now available directly to consumers in

---

[1]   The concept of the "electronic manuscript" is closely related to another idea: *electronic publishing* (cf. Kleper, 1990). Kist (1988) describes "electronic publishing" as "the application of computer-supported procedures by a publisher in such a way that information is retrieved, assembled, organized, stored, updated, and, using various data-exchange systems, distributed in a wide variety of forms to the appropriate parties". A need has even developed for an abstracting service devoted exclusively to electronic publishing: *Electronic Publishing Abstracts*, issued monthly by Pergamon Press and restricted to summaries of the latest articles dealing with electronic publishing and related information technology.

[2]   D. Seebach, chairman of the Advisory Board of the journal *Chimia*, observed in a lead article (*Chimia* **1988**, *42*, 122) entitled "Revolution in Publication Technology from the Eyes of a Consumer": "I am convinced that publications in the field of chemistry will in the near future make their way from the author, through the editorial staff, and finally to the printer via a series of compatible data-processing systems. Discussions along these lines are now taking place in every editorial office. One consequence will be the transfer of certain activities from editors, managers, and printers back to the office of the author". It is interesting to note in this context that a recent major review article by Seebach ("Organic Synthesis—Where Now?" *Angew. Chem. Int. Ed. Engl.* **1990**, *29*, 1320; *Angew. Chem.* **1990**, *102*, 1363) was in fact prepared and processed in precisely this way—both text and figures—including translation from German into English.

the form of diskettes or compact discs (CD-ROM packages; cf. Sec. 5.4) as an alternative to printed books or periodicals. In other words, the front lines of the electronic document revolution have advanced all the way to the public marketplace. Long-distance transfer of information via data-transmission networks and telephone lines is also becoming increasingly routine, providing people everywhere with instantaneous access to huge centralized databases subject to regular updating with the latest information. For example, the most efficient way to conduct an exhaustive scientific literature search today is to tap into such resources as the "online" versions of *Chemical Abstracts* or *Science Citation Index*, a prospect few could have envisioned twenty years ago. Several publishers also offer complete online versions of their scientific journals—which brings us back to our real focus of attention: electronic manuscripts.

As we shall see later, electronic manuscripts can originate in various ways, and they exhibit an even wider variety of forms. The simplest is the pure *ASCII file* (*text file*, or *text-only file*; cf. Sec. 1.1.2), or an ASCII file that has been embellished with a few cryptic formatting instructions (*codes*, leading to a *coded manuscript*; cf. Sections 1.1.3 and 1.1.4). A more complex entity is the electronic manuscript that results from saving a document according to the dictates of a text-editing (word-processing) program, since such files always include a host of (invisible) formatting commands meaningful only within the context of that particular program (cf. Sections 1.1.5 and 9.6).[3]

The most complicated type of electronic manuscript from the author's perspective is one prepared so meticulously—usually with a combination of word-processing and page-makeup software—that a publisher can use it to generate the final version of a document simply by outputting the file to a *laser typesetter* (or *imagesetter*; Sec. 7.5). Closely related is the carefully crafted document printed by the author with a high-quality laser printer, leading to copy that the publisher can treat as "camera-ready". In this case a straightforward photographic process becomes the key to transforming the author's work into a finished product (cf. Sec. 1.1.6).

Manuscripts of the latter type are rapidly becoming of much more than theoretical interest for practicing scientists. Publishers everywhere are finding it increasingly difficult to cope with harsh economic realities. Technical

---

[3] ASCII manuscripts (Sec. 1.1.2), coded documents (Sections 1.1.3–1.1.4), and word-processor files (Sec. 1.1.5) are treated here as though they were three distinct species, but an infinite variety of combinations and compromises can also be imagined, and these in fact tend to be the rule.

monographs in particular are relatively expensive to produce, and their preparation constitutes a serious financial risk, but the author able to present for publication a manuscript in the form of high-quality camera-ready copy is in an unusually strong bargaining position. Our book is dedicated to encouraging authors to acquire precisely the skills required to provide such manuscripts.

In the sections that follow we examine much more closely various types of electronic manuscripts, considering them through the eyes not only of the author but also of the publisher.

1.1.2 ASCII Manuscripts

As suggested above, the most primitive type of electronic manuscript is the "ASCII file". "ASCII" is an acronym for *American Standard Code for Information Interchange*, a "7-bit" information system that establishes unique numerical equivalents for $2^7 = 128$ different typographic symbols. Thus, specific numerical values have been assigned to each letter of the alphabet (a, b, ..., A, B, ...), each numeral (0, 1, ..., 9), several punctuation marks (. , ; ! and others), and certain special symbols (such as + and %), together with a few crucial control functions (i.e., "line feed", "carriage return"). As its name implies, the ASCII code was developed in the United States, but it has been adopted worldwide as a common basis for communication within and among a wide variety of general-purpose data-processing devices.[4] The first 32 of the 128 ASCII representations (codes 0–31) are reserved as *control codes*, intended mainly to facilitate data transfer to printers and other peripheral devices. A few representative ASCII (decimal) codes together with the corresponding symbols and functions are listed in Table 1–1.

In the following discussion we will regard an *ASCII* (or *text-only*) *manuscript* as one that consists exclusively of ASCII-code representations[5] of the

---

[4]   A less widespread alternative to ASCII is the 8-bit EBCDIC (*Extended Binary Coded Decimal Interchange Code*) system, with $2^8 = 256$ equivalencies.

[5]   The definition of an ASCII text file should perhaps be broadened slightly to include a few additional language-specific linguistic symbols (all European), such as the German umlauts (ä, Ö, ...) and accented letters like é and ø. These are provided for in an *extended ASCII code*, with the supplemental code numbers 128–255, but individual computer manufactures (Apple, Atari, IBM, etc.) have resisted uniform adoption of certain aspects of the extended code.

Table 1–1. Examples of ASCII code assignments.

Code	Significance	Code	Significance
2	STX (start of text)	43	+
4	EOT (end of transmission)	48	0
10	LF (line feed)	49	1
13	CR (carriage return)	65	A
33	!	66	B
35	#	97	a
40	(	98	b

letters, numbers, and symbols constituting the text of a particular document. This restriction implies that virtually the only commands available for providing *structure* to the document (e.g., headings, beginnings of paragraphs, and other aspects of formatting) are "carriage return" and "line feed". It is certainly possible in principle to incorporate other, more powerful instructions—in an encoded form, perhaps—but everyone with access to the document in question would need to agree on an unambiguous set of conventions (cf. Sec. 1.1.3).

In general, formulas, tables, and illustrations must be excluded from the ASCII version of a document, because they entail too much special formatting. This is one of several reasons why publishers insist that electronic manuscripts always be accompanied by complete printed copies of the documents. An editor can then assume responsibility for carefully annotating ("marking up"; cf. Sec. 3.2) the "hardcopy" version, designating in a specific way any text that is to be set in *italic* or **boldface** type, for example. This often involves use of an elaborate system of underlining, with different colors sometimes employed to indicate different types of highlighting (see Sec. 2.3 and Chap. 3 for information regarding the significance of different type styles). Explicit markings are also added at this time to alert the typesetter regarding approximate places where footnotes, figures, tables, or formulas should be inserted. In most cases the author is expected to provide the publisher only with reasonable *models* of required supplementary elements (e.g., figures, tables, graphs). These also must be annotated by a trained editor, and sometimes extensively modified. Graphic parts may even need to be redrawn professionally before the entire package can be passed along to a typesetter, who assumes responsibility for merging the various elements into a single printed document.

Most electronic text is created today with word-processing software, resulting in *formatted* data files (Sec. 1.1.5) that are considerably more comple than ASCII files. Nevertheless, word-processing programs usually provide an option for saving data in "text-only" ASCII form. This often overlooked feature generally represents the easiest way of stripping out program-specific control codes that might otherwise seriously impede automated typesetting.

Simple ASCII files continue to be very popular with publishers, in part because they present authors with few opportunities for causing problems.[6] *Reformatting* a complex electronic document can be a very time-consuming process. Indeed, it is almost always easier for a publisher to deal with a file containing too little formatting information than one that is filled with inconsistencies or what the editor regards as inappropriate elements (unconventional mathematical or chemical formulas, for example, or literature citations or footnotes presented in unacceptable ways).

1.1.3 Coded Manuscripts

In recent years many technical journals, such as the *Journal of the American Chemical Society* and *Angewandte Chemie*, have begun to experiment with and even solicit electronic manuscripts submitted on diskettes. A few journals (e.g., the *Journal of Physical Chemistry*) go a step farther and offer special facilities for receiving electronic manuscripts via data-transmission networks (cf. Sec. 4.5).[7] Unfortunately, however, no generally accepted standards yet

---

[6]   Publishers are in general agreement that, except in the case of SGML files (Sec. 1.1.4), the more sparingly a manuscript is coded, the easier it is to refine and prepare for publication. In particular, no attempt should be made by the author of an ordinary electronic manuscript to impose design specifications on the text. One possible exception is use of a system for designating the *hierarchical levels* of headings, and it may be permissible to indicate characters that are to be set in special type. Text associated with footnotes, tables, and figure captions must normally be consigned to separate data files, and graphic elements such as illustrations and formulas are often accepted only in "hardcopy" (printed, hand-drawn) form.

[7]   Experiments have even been undertaken with completely "electronic journals", distributed to the end users by electronic means. For a thoughtful discussion of this phenomenon and some of its implications, see S. Borman, "Advances in Electronic Publishing Herald Changes for Scientists", *Chem. Eng. News*, 14 June 1993, pp. 10–24.

Table 1–2. Codes recommended in the *Chicago Guide to Preparing Electronic Manuscripts*.

Letter or symbol	Code	Interpretation
β	`<beta>`	Greek letter β, normal type
–	`<minus>`	minus sign in a mathematical equation
>	`<gt>`	"greater than" (as distinct from the symbol ">" itself, which is used as a "code delimiter")
Â	`<cir>A`	circumflex accent over capital "a"
ä	`<um>a`	umlaut over small "a"
	`<i>`	start setting in italic type
	`</i>`	stop setting in italic type (return to normal)
a^{12}	`a<sup>1<sup>2`	superscript "12" directly following "a"

exist regarding the form such manuscripts should take (though one important step in this direction is discussed in Sec. 1.1.4). Each journal instead prepares and distributes its own set of "Instructions for Authors", often supplemented with an extensive list of codes for labeling key features in the text (e.g., the start of a new paragraph, a third-order heading, a line-break essential for the insertion of an equation, or characters that are to be set in italic type).

One of the early attempts to introduce order into this chaotic situation was publication of the *Chicago Guide to Preparing Electronic Manuscripts for Authors and Publishers* (in 1983, revised in 1987, as part of the series *Chicago Guides to Writing, Editing, and Publishing*). The editors of this little spiral-bound volume were interested in sharing with a wide range of authors (and other publishers!) one particular publisher's insights into the many advantages to be gained by encouraging authors to prepare electronic manuscripts—but also some of the associated pitfalls, especially in the absence of universal standards. Included was the specific set of typesetting codes that had evolved over the years at the University of Chicago Press, a few of which are illustrated in Table 1–2.

The extra burden imposed upon an author/scientist undertaking to prepare a "coded manuscript" of the type described quickly becomes apparent from a single example. Thus, it takes the long, complex string of characters

```
<eq><i>x</i> = exp (<i>a</i> <minus> <i>b</i>)</eq>
```

to express in a manuscript based on the "Chicago code" the relatively simple mathematical expression

$$x = \exp (a - b)$$

containing the three italicized variables *x*, *a*, and *b* (cf. Sec. 3.4). Special code markers ("`<eq>`" and "`</eq>`" for "equation", "`<i>`" and "`</i>`" for italics) have been introduced to indicate the beginning and end of each element that must be typeset in a special way. Coding is also required in this case to specify the minus sign, since this is not the same as a hyphen (cf. Sec. 3.3), and it has no ASCII equivalent. Codes in the Chicago system are essentially mnemonic, and they are set off from surrounding text by angular brackets, a convention adhered to by other coding systems as well (cf. Sec. 1.1.4). Professional typesetters have long been accustomed to working with "tags" of a similar sort, although theirs tend to be much more complicated (cf. the descriptions of T$_E$X and JUSTTEXT in Sec. 10.5).

Not surprisingly, authors have shown relatively little enthusiasm about assuming responsibility for embellishing their manuscripts with such arcane symbolism, particularly authors of long manuscripts. Nevertheless, the concept of coding is an important one for reasons that will become more apparent in the section that follows. Fortunately, special software (or even an appropriate set of "macros") can simplify the coding process to the point of making it quite manageable.

Mathematical formulas and equations are particularly awkward to incorporate into electronic manuscripts because of their structural complexity (cf. also Chap. 3). The *Chicago Guide* strongly recommends investing in and learning to use special software created explicitly for this purpose, such as a program based on the language T$_E$X (Sec. 10.5), in part because the resulting codes can often be understood by professional typesetting equipment. Many publishers disagree, however, insisting that authors include in their manuscripts only sequential numbers to *represent* required mathematical or chemical formulas, providing as an appendix to the manuscript a correspondingly numbered set of printed or hand-written formulas for professional processing by experts.[8]

---

[8]   Authors wishing to incorporate professional-looking formulas directly into the text of *camera-ready* manuscripts (Sec. 1.1.6) can take advantage of several powerful and convenient formula-building accessory programs, a few of which are described in Section 9.7.

1.1.4 Standardized Coding: SGML

Despite all their advantages, electronic manuscripts present a publisher with serious challenges. For example, it is not always easy to establish full *data compatibility* between various computer systems,[9] a problem that initially stimulated demand for sophisticated *conversion software* used in conjunction with appropriate supplementary hardware, such as alternative disk drives.

Once the full scope of the compatibility problem became apparent, however, the printing industry worldwide began to devote serious efforts to the development of unified guidelines that could be used for the coding of electronic manuscripts prepared in ASCII form. The goal was to devise a universal system that would permit files from a typical author's personal computer to drive phototypesetting equipment like that found at most printing plants, thereby eliminating the need for extensive "retyping" and at the same time simplifying the task of formatting. A major incentive behind this effort was the possibility of avoiding the many mistakes that inevitably creep into text that has been recreated ("keyed in") manually by someone other than the author. After much study and extensive negotiations an international coding standard *(ISO 8879-1986)* was finally adopted based on a document language called *SGML (Standard Generalized Markup Language).*[10]

The ISO norm itself provides only the *structural scheme* according to which documents and their constituent *text elements* (logical subunits: headings, chapters, sections, footnotes, tables, etc.) are to be coded, not the actual *symbols* to be used for that coding. Thus, each distinct text element is

---

[9] Peters (1988) observed cynically that "desktop publishing is the penalty typesetters brought upon themselves by failing to come to terms with data conversion". For more information on problems of data conversion see Card (1990).

[10] The Association of American Publishers earlier formulated its own *Standard for Manuscript Preparation and Markup*, the *AAP Standard*, also on the basis of SGML. Since 1988 SGML has enjoyed special priority status within the United States government under the guise of "FIPS", the *Federal Information Processing Standard*, and other countries have adopted similar guidelines. Certain national groups have even gone so far as to provide support software to assist authors in the preparation of "standard" electronic manuscripts (e.g., the German program *strukTEXT* from the Bundesverband Druck e.V., Wiesbaden, and the Börsenverein des Deutschen Buchhandels e.V., Frankfurt). William Tunnicliffe of the Courier Corporation is credited with first promoting the idea of "generic coding", which underlies the SGML concept, in a presentation to the Canadian Government Printing Office as early as 1967.

to be assigned a characteristic "tag" to identify it according to the structural role played by its content (*content-based coding, generic coding, generic markup*; authors sometimes refer to this as *data-based publishing*).

SGML coding operates on the premise that the key to describing a document is analyzing it in terms of its component parts: recognizing a particular footnote, for example, as one discrete *element* within a larger entity, an element subject to independent manipulation even though it must eventually be incorporated—somewhere, and with a yet-to-be-determined typeface, line width, and line spacing—into the overall document. For coding purposes this footnote can be regarded simply as one of many arbitrary, isolated sets of character symbols, albeit one that must retain a firm *link* to a specific point in the main body of the text. All the various elements that constitute a given document must be singled out in this way and then carefully labeled as entities of particular types. Once this has been accomplished the important further assumption is made that all elements with equivalent labels will also share a common set of physical characteristics, and a common mode of structural connection with the document as a whole. This being the case, it becomes possible to defer decisions with respect to a *particular* set of typographical characteristics (margins, typefaces, etc.) to a later stage in the publishing process.

SGML coding achieves its fundamental purpose very effectively, in that it greatly simplifies the process of eventually supplying formatting and typesetting instructions, and it ensures complete flexibility in this regard. Formatting parameters can also be easily changed to reflect changed circumstances, as in the need to prepare a new edition of a book with drastically altered page dimensions, perhaps, or a CD-ROM version of a printed work. Moreover, an SGML-coded document can readily be integrated into a larger database containing other similar documents, all of which then become subject to selective search-and-retrieval operations sensitive not only to content but also content as a function of constituent elements. Indeed, the adoption of SGML standards can dramatically reduce the effort—and cost—involved in preparing a document for typesetting, as well as in the actual typesetting process.[11]

---

[11] Advocates of SGML believe very strongly that electronic manuscripts *themselves* are valuable entities, and that they deserve to be prepared in such a way that they can be preserved *(archived)* and exploited indefinitely, and with maximum flexibility. The ideal electronic document storage system would be both timeless and completely independent of particular word processors or typesetting devices—and

As with the *Chicago Guide* system, SGML tags usually take the form of mnemonic abbreviations set in angular brackets, often of a fairly self-explanatory nature (e.g., "FN" for "footnote"). Properly coded SGML files can be introduced directly and in their entirety into suitably programmed typesetting machines, and such files may even contain tables and certain types of formulas. Compatibility at a level this high is rarely achieved with documents prepared by standard word-processing techniques.

Several "user-friendly" computer programs *(SGML editors)* are now available to assist authors in the preparation of SGML-based electronic documents; examples include SOFTQUAD AUTHOR/EDITOR and WRITERSTATION.[12] Their main purpose is to facilitate the coding process, at the same time ensuring consistent adherence to SGML guidelines. The user is offered convenient access, for example, to all the various SGML tags, eliminating the need for introducing them manually. Resulting text can be displayed on the screen (or printed) either with or without the embedded codes, and sophisticated tests automatically check for consistency and plausibility of coding. Figure 1–1 shows a typical screen display from one such system, SOFTQUAD AUTHOR/EDITOR, the latest version of which (Version 2, released in 1993) even provides sophisticated features for easy construction of tables.

A generically coded manuscript may be less satisfying visually to its author than an attractively formatted finished product,[13] and preparing it is likely always to be somewhat of a nuisance, but at least the SGML approach is one to which authors can easily relate, since it does not impose the jargon or even the perspective of the typesetter. Thus, labels introduced into an SGML

---

certainly independent of such ephemeral characteristics as page size and typeface. At the same time such a system would also address very diverse concerns applicable to authors, editors, and typesetters, all of whom are becoming increasingly dependent on computer-driven tools. Certain aspects of this problem are the subject of a short communication by H. F. Ebel, C. Bliefert, and W. E. Russey in *European Science Editing*, May 1993 (No. 49), pp. 12–13.

[12] SOFTQUAD AUTHOR/EDITOR is distributed by SoftQuad, Inc., and is available for a wide range of computers (e.g., MACINTOSH, IBM, and UNIX-based systems), with full cross-platform compatibility. WRITERSTATION, from Datalogics, Inc., is restricted to IBM-compatibles and VAX systems (see Appendix C for addresses).

[13] It should be noted, however, that SOFTQUAD AUTHOR/EDITOR does permit an author to assign *temporary* formatting to the various elements of a document, which in turn makes it possible to print at will attractive single copies—and to view a document on the screen in "WYSIWYG" form (cf. footnote 19 in Sec. 1.2, as well as Sec. 6.3).

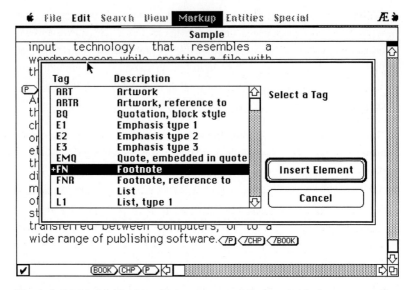

Fig. 1–1. Dialog box for introducing a new text element (with the corresponding SGML "tags") into a document created with SOFTQUAD AUTHOR/EDITOR.

manuscript simply *identify* features like "highlight" or "footnote" or "1st order heading" rather than conferring upon them their technical specifications, such as "12/14 Times roman × 20, flush and hang 1 em".[14]

Authors can also expect to derive certain tangible benefits from adopting the SGML system:

● The system imposes a high degree of order upon one's work, since all text elements must necessarily be declared and interrelated in a consistent and hierarchical way, starting with the highest-level element "document". In the words of SOFTQUAD promotional literature: "[SGML] gives data a base … [and] information an architecture". Moreover, SGML software helps the author remain always alert to the precise role of the text currently being created: the fact that it is body text, for example, or a second-order heading, a footnote, or a figure caption.

---

[14] Times roman font in 12-point size with 14-point leading and set on fully-justified 20-pica lines, the first line of which is indented by 1 em (cf. Chap. 2).

- SGML authors are rewarded with very elegant facilities for keeping track of citations, and preparation of a table of contents or an index can be a remarkably straightforward process with an SGML-coded document.

- The presence of structural tags actually facilitates some aspects of manuscript revision. For example, to transform a bibliographic citation into a footnote one might simply change its tag from "BB" to "FN".

- "Search" and "replace" operations can be narrowed to individual structural categories (footnotes, for example, or figure captions, or highlighted text).

- Actual formatting details (type size, line spacing, etc.) need be specified only once for an entire manuscript regardless of its length—and these specifications can be added by an expert at the publishing house, someone with considerable experience in conferring form upon printed works, and in choosing parameters that are most likely to achieve a desired effect.[15]

Several publishers have expressed a willingness to make SGML software available to their authors, together with all the instructions necessary for its use, but authors for their part tend to resist experimenting with radically new working techniques. It seems in fact unlikely that SGML will achieve a truly major breakthrough unless:

- publishers or third parties assume the responsibility for converting authors' conventional electronic manuscripts into SGML form, or

- software developments make it exceptionally convenient for authors to undertake the necessary conversions themselves.

Further insights regarding these issues—and SGML generally—are provided by Bryan (1988), Goldfarb (1990), Kist (1988), Smith (1986a, 1986b, 1987), and van Herwijnen (1990).

---

[15] Some of the advantages of generic coding are also reflected in the "style sheets" that are now a part of many word-processing and page-layout programs (cf. Sections 9.4.1 and 10.2).

1.1.5 Manuscripts Prepared with Standard Word Processors

As suggested earlier, file conversion problems can impose formidable barriers for the publisher anxious to process authors' text files that have been created with ordinary word-processing software. Especially in the case of a document containing tables, or with paragraphs formatted in unusual ways, program-specific embedded coding may become extremely complex, which can lead to serious errors in translation. Footnote commands are notorious sources of trouble. For this reason it is absolutely essential that authors come to a clear understanding with prospective publishers or printing houses at an early stage in manuscript development with respect to what types of word-processing files are acceptable, and what limitations, if any, may apply. In many cases an author will be asked to submit a sample file containing a representative selection of proposed text elements so that definitive printing tests can be conducted.

Considerations relevant to the selection and proper application of word-processing software are the focus of Chapter 9.

1.1.6 Camera-Ready Manuscripts

In the preceding discussion it has been taken almost for granted that at some point prior to final typesetting extensive intervention will be required on the part of a professional editor. For example, it is not at all unusual for an editor to discover that an author has incorporated into an electronic document satisfactory instructions for certain types of formatting (e.g., italicization), but has done so inconsistently (ignoring, for example, essential italics in literature citations, or in mathematical equations). Other important aspects of formatting may have been overlooked completely, such as distinctions between headings at various hierarchical levels. Dealing with mechanical problems such as these has long been an obligatory—albeit unpleasant—part of an editor's daily routine. As time passes, however, more authors are beginning to accept the challenge of preparing *fully formatted* copy suitable for direct publication without such editorial intervention. Some scholarly journals (e.g., *Tetrahedron Letters*) even insist upon receiving "camera-ready" copy, ostensibly as a way of avoiding the delays (but also the costs!) that are necessarily associated with formal typesetting. This provides authors with a powerful incentive to learn something about working effectively with powerful word processors and page-layout software—as well as high-quality laser printers—and to abandon the venerable typewriter.

It should be emphasized from the outset, however, that the effort involved in preparing a *truly* high-quality, camera-ready manuscript, especially for an entire book, is far greater than most authors (or translators) anticipate. For all the hymns of praise dedicated to "Desktop Publishing", the author or translator with sufficient patience to develop a complete camera-ready book manuscript remains a rarity.

Generating camera-ready copy obviously requires access to and familiarity with a computer and an appropriate range of software. This is only the beginning, however. Considerably more is required, including experience, a certain amount of talent, and enough motivation to be willing to invest long hours in ensuring the use throughout of appropriate fonts (of the right size), proper line and character spacing, layouts that achieve a balanced text density—in short, all the characteristics one takes for granted in a first-class printed document. Beyond that one must also skillfully incorporate the requisite illustrations, diagrams, and other graphic elements, all in an efficient but attractive way. The quality of the final product will inevitably bear the stamp of qualities inherent in the practitioner.

Preparing camera-ready copy also presupposes a willingness to adhere strictly to layout and design criteria that are acceptable to the publisher. Extensive and frank dialog should be initiated at an early stage with both the editorial office and the production department at the publishing house. In preparation for these exchanges the prospective author of a camera-ready manuscript should assemble several (!) well-chosen test pages, each carefully prepared according to guidelines furnished by the publisher. Examples should be provided of every feature that can reasonably be expected to appear anywhere within the work: text set in a variety of type sizes and styles, for example, all with appropriate spacing, as well as figures, tables, footnotes, formulas, equations, and any other "special" elements, such as marginal notes. Considerable discussion and negotiation is certain to follow, related to such technicalities as optimal type size in various contexts, the all-important index and how it can best be prepared, possible applications of color, idiosyncrasies associated with the proposed production process, and many others.

This book has been written primarily to acquaint practicing scientists and engineers with some of the most effective techniques available for preparing camera-ready copy of reports, professional communications, technical articles, and especially books. While the decision to become a "desktop publisher" (cf. Sec. 1.3) should never be taken lightly, the number of authors intrigued by the challenge of preparing finished versions of their own publications "at home, in the study" is clearly growing. This development is almost certainly a desirable one from the standpoints of both scientific publishers and the scientific community at large, but the parties most directly involved—authors and publishers—must learn to approach proposed desktop-publishing ventures with due caution. Inexperienced author/typesetters almost always require a considerable amount of help from professionals, including editors and production personnel. This is true even with respect to relatively straightforward, unembellished text, especially if the finished product is to reflect a truly "professional touch". "Scientific" text is further complicated by the presence of such potential hazards as equations and graphs.

Anyone proposing to embark on the path of a "desktop publisher" should begin by pondering the following observations:

- One absolute prerequisite is convenient and extensive access to (and thorough familiarity with!) all the requisite hardware and software. Despite frequent protestations to the contrary, especially in advertisements, most people find that it takes many months to accumulate the knowledge and skills required for turning out consistently good results. Publishers have good reason for asking prospective typesetter–authors to supply samples of their previous work, for only on this basis can they judge one's real ability to provide a high-quality finished product within a reasonable period of time.

- Few scientist–authors have any formal training in typography. This implies a need for explicit *typesetting guidelines*, complete with appropriate examples presented in a scientific context. Such guidelines must not be formulated in the technical terminology of the print shop, but should instead feature explicit, simplified explanations and illustrations, analogous to those in Chapters 2 and 3 of this book.

- Publishers often seek early assurance that *all* changes they may later request will in fact be conscientiously incorporated by the author. Needless to say, most authors tend to resist introducing into their masterpieces "improvements" dictated by others.

- The editorial staff member who is to supervise the work must be fully informed at the outset regarding any limitations that may be associated with the author's computer hardware or software.

- Editors sometimes find it necessary to suggest major revisions in a manuscript, and the danger is real that at some point the developer of a camera-ready manuscript may begin to function more like a professional typesetter or graphic artist than a creative writer, paying greater attention to technicalities of type style and line spacing than to the text itself. Experienced editors shudder at the thought of the problems this can create!

1.2 Milestones in the Historical Development of "Desktop Publishing"

It is a remarkable fact that it is now entirely feasible for an author equipped with little more than a simple personal computer to undertake the preparation of true camera-ready copy

- at relatively low cost
- with a minimal amount of technical training

Several key developments in computer technology are responsible for this development (cf. also Straka et al., 1987, or Kleper, 1990). One milestone was Canon's perfection in 1983 of the "engine" that would form the heart of low-cost, tabletop laser printers, a direct outgrowth of the company's experience with photocopiers. Hewlett-Packard was the first company to market (in 1984) a complete low-cost printer built around the Canon mechanism, the HP LASERJET. This in turn provided personal computer users with the ability (at least in principle) to turn out relatively high-quality printed copy—incomparably better than that associated with the average impact dot-matrix printer—and to do so inexpensively.[16] The next important step was achiev-

---

[16] The quality of laser-printer output (usually 300 dpi, cf. Sec. 6.2) still falls well short of the demands of many book publishers, who are accustomed to phototypesetting machines and laser typesetters (cf. Sec. 7.5) operating at resolutions in excess of 1500 dpi.

ing compatibility between such laser printers and a wide variety of sophisticated text and graphics programs: a standard *software interface*—in particular, development of the page-description language PostScript (cf. Sections 7.3 and 7.6).

Creating a "friendly" *user interface* for the computer, incorporating a *graphics-based* screen display, also played a crucial role (Sec. 4.2). Features generally associated with such an interface include:

● *windows* for selectively displaying text, data, graphics, etc.

● *pulldown menus* for selecting any of several applicable program functions[17]

● *icons* (pictograms) to assist in the rapid visual identification of specific programs (cf. Fig. 1–2) and program functions (Fig. 1–3)

● the convenient input device known as a *mouse*, which facilitates visually guided selection of a particular program function from among several listed in a menu or dialog box, or the accurate positioning of an on-screen cursor, all without recourse to the keyboard (Sec. 4.4)

Features like these are important because they free the user to concentrate directly on creative activities rather than on the computer and its software. No longer is it necessary for an author to conform rigorously to a computer programmer's ways of thinking and communicating in order to perform a simple task like copying a file to a particular diskette, or executing a routine sequence of data-manipulation steps. The result has been a dramatic simplification of the learning process.[18]

Another key factor in the growth of desktop publishing has been the increasing availability—at decreasing cost—of computer *memory* (cf. Sec. 4.2). Only with this development (clearly foreseeable by about 1984) did the graphics-based user interface become truly feasible, opening the door to a whole new category of users. Expanded memory also made it practical for

---

[17] The presence of such menus is usually signaled by a set of largely self-explanatory titles in a display panel at the top of the screen, each of which can be "rolled down" like a window shade to reveal its contents on command.

[18] In fairness it should be noted, however, that user-friendly software is also highly memory-intensive, and it tends to be relatively slow, all of which makes it less well suited to certain professional applications—but that was never intended to be its role.

Fig. 1–2. Examples of icons used for designating various application programs, displayed here on a MACINTOSH "desktop" (cf. Sec. 4.2). The "reversed" colors of the lower-left icon indicate that this one has been "selected".

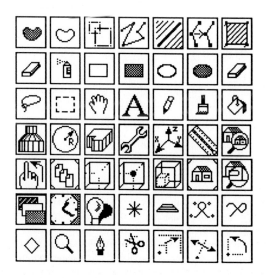

Fig. 1–3. Icons representing control functions within specific MACINTOSH programs.

anyone to create and freely manipulate memory-intensive elements like complex illustrations, which are subject to placement directly alongside conventional text, all within the confines of an economical personal computer. Finally, increased memory facilitated real-time screen display of a wide variety of type styles and type sizes, providing a close approximation of the document that would ultimately be printed—a highly desirable feature, though admittedly one that is not absolutely essential, especially for the author concerned only with straightforward text.[19]

1.3 The Term "Desktop Publishing"

"Desktop publishing" has become something of a cliché in recent years,[20] as computers have relentlessly displaced the typewriter and new generations of printers have made it possible to prepare in one's own office printed copy that compares favorably from an esthetic standpoint with the work of a professional.

It will be useful to begin our formal discussion of the subject with a few observations regarding the seminal term itself: *desktop publishing* (abbreviated DTP), a phrase apparently first introduced by Paul Brainerd, president of the Aldus Corporation. Unfortunately, some of the images the phrase conjures up can be misleading, and it has clearly been responsible for a certain amount of misunderstanding and even outright hostility.[21] The significance in the present context of the word "desktop" is fairly obvious;

---

[19] This is perhaps an appropriate place to introduce the delightful term "WYSIWYG", an acronym for "What You See Is What You Get", often used to describe software capable of displaying on the screen "exactly" what will later emerge from a printer. We will have quite a bit more to say about this subject in Section 6.3.

[20] Related terms include "computer-aided publishing" (CAP), "computer-integrated publishing" (CIP), "desktop electronic publishing" (DEP; Marshall, 1990), "electronic technical publishing (ETP), "computer publishing", "electronic publishing", and "computer-aided writing" (CAW; Ebel, Bliefert, and Russey, 1987).

[21] It is for this reason that we have chosen to introduce the term initially in quotation marks. The French have prudently adopted more accurate descriptions: «publication assistée par ordinateur» (PAO: "computer-assisted publication") and, even more to the point, «mise en page à l'écran», ("typesetting on the screen").

it is meant to suggest a process carried out conveniently at one's own workplace.[22] The problem arises with the second word: "publishing". The verb "to publish" has a number of very diverse meanings, including *arranging for* the publication of an article in a journal, but its most fundamental purpose is to express the sum of all the activities involved in transforming a manuscript into a widely disseminated ("public") product, including typesetting, printing, binding, and distribution. Various treatises on the subject of desktop publishing employ definitions such as

- "the combining of text and graphics into high-quality publications" (Marshall, 1990)

- "electronic page makeup on personal computers" (Bove, Rhodes, and Thomas, 1986)

- "the facility to generate and combine text and graphics by electronic manipulation into a series of pages which are output through a laser printer" (Hewson, 1988)

- "designing and producing pages on a microcomputer" (Gosney and Dayton, 1990)

In other words, what is usually meant in the "desktop publishing" context is *not* publication (of a book, say) in the broad sense—and most certainly not the chores of binding and distribution, except perhaps in the case of a work limited to a very few copies. Even the printing process would normally be excluded, apart from preparation of a few sample copies.[23] Instead, the real goal is to create in one's home or office the best possible *model* of a complete printed work, perhaps including graphic images, for subsequent reproduction elsewhere, avoiding in the process the cost and inconvenience of a

---

[22] Similar reference is made to other "desktop" activities: "desktop communication", "desktop design", "desktop electronic productivity", "desktop engineering", "desktop mapping", "desktop presentation" (cf. Sec. 1.6), "desktop productivity", and "desktop video". A different use of the word relies on the "desktop" as a way of characterizing a graphic screen simulation of a work surface, one facet of a user-friendly software interface. Thus, Apple calls the primary screen display of its MACINTOSH line of computers a "desktop", equipped with icons depicting such handy office "tools" as file folders, applications programs, and even a wastebasket.

[23] Depending upon the number of pages involved, one might conceivably produce 2, 3—even 10—original copies with a laser printer directly, but it would be sheer folly from the standpoint of both cost and time to use a laser printer to prepare hundreds of copies of a document for subsequent binding and wider distribution.

complicated and expensive typesetting operation. One descriptive brochure from Hewlett-Packard describes DTP in terms of "PC-supported" creation of high-quality printed output with the aid of a laser printer. Other restrictive definitions have relied on phrases like "desktop typesetting", "a layout table on the screen", "laser typesetting", "use of computer type fonts", or—more to the point—"the computer-supported manipulation of text and graphics".

One almost unlimited market for the application of DTP techniques is the business community, with its constant demand for such documents as product brochures, corporate newsletters, and operating instructions.[24] Some of the latter may indeed be designed for consumption by at least a segment of the public, but they still would not class as "publications" in the strict sense of the word. Most desktop publishing in the new "era of typographic data systems" is really only a highly refined form of word processing—but it has certainly become one of the most important applications yet devised for the personal computer.

A revealing measure of the current significance of desktop publishing—however it might be defined—is the amount of space devoted to the subject in computer magazines. Another is the number of periodicals devoted almost exclusively to DTP and its various applications, such as *Publish*, *PC Publishing and Presentation*, *microPublishing Report*, *InfoWorld*, and *Personal Publishing*. The titles of countless books in English (and even foreign languages!) combine the fashionable word "desktop" with others such as "typesetting", "layout", or "typography" in further confirmation of the considerable impact of this particular application of data-processing technology; e.g.:

> Baumann and Klein (1989), Blatner (1991), Bove et al. (1986), Bradley (1992), Brown (1989), Busche and Glenn (1991), Cavuoto and Beck (1992), Cole and Odenwald (1989), Cookman (1993), Glover (1989), Günder (1988), Heck (1992), Hewson (1988), Jones (1987), Kleper (1990), Marshall (1990), McClelland and Danuloff (1987), Menousos and Tilden (1990), Peters (1988), Sitarz (1989), White (1988), Yasui (1989), Ziegfeld and Tarp (1989).

One can readily distinguish three discrete categories of activity within the overall field of desktop publishing—which we will henceforth regard as *the electronic preparation of copy for subsequent duplication*:

---

[24] For more information regarding the corporate role of desktop publishing, directed especially to management, see the book by Marshall (1990).

● pure text preparation

● the creation and manipulation of graphic images, where text plays only a minor role

● complex page makeup, in which text and graphic elements are united in a harmonious way within the confines of a single page

Some of the steps involved in these three types of activity are illustrated schematically in Fig. 1–4, although in practice the three categories frequently overlap.

Much desktop publishing work can be accomplished with a relatively inexpensive computer system and a single piece of advanced word-processing software (cf. Chap. 9). Indeed, "high-end" word-processing programs have evolved to such an extent that they increasingly resemble what would once have been called page-makeup software.[25] At the same time, programs originally designed primarily for page makeup are incorporating more and more features to facilitate text editing.[26]

Nevertheless, the ideal environment for working conveniently with both text and graphics always includes a *package* of programs offering a wide range of specialized features. This in turn presupposes a high degree of *compatibility* between the various software elements, permitting data originating in one program to be transferred easily and reliably into another. For example, a good makeup program today would be expected to handle conveniently not only formatted text from a number of sources, but also illustrations, graphs, formulas, and the like, which may have been created or modified with the aid of several different programs. A decisive role in the process of transferring information from one program environment to another is played by the hardware on which the software "runs". The Apple MACINTOSH line provides an excellent example of a DTP system that im-

---

[25] A recent review of the MACINTOSH programs WORD, RAGTIME, and PAGE-MAKER described all three as "DTP programs with specialized layout features" (cf., for example, the "Print Preview" and "Page Layout" functions in WORD, described in Sec. 9.4.1).

[26] Originally, layout or makeup programs were regarded strictly as "enrichment software", since early versions concentrated exclusively on the *placement* of existing text and graphics on a page, with few provisions for editing. This is certainly no longer the case; editing features in the current version of PAGEMAKER, for example, rival (or even surpass) those of a word-processor like WORD (cf. Sec. 10.2).

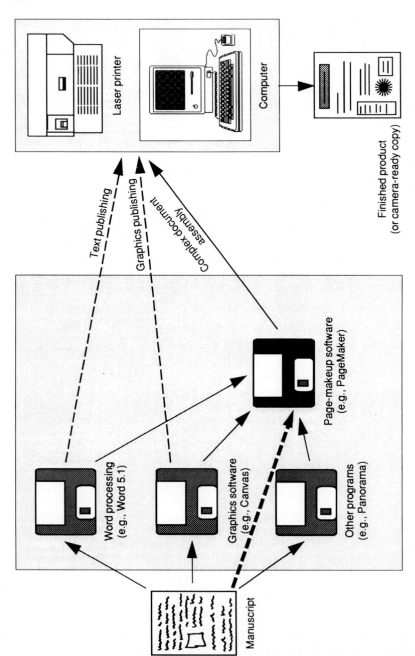

Fig. 1-4. Schematic diagram of activities involved in transforming different types of manuscript copy into a complex document by the "desktop publishing" method.

poses virtually no obstacles to software or hardware compatibility. This particular system is discussed in considerable detail in Sections 4.2 and 4.4.

Returning to the three general classes of desktop-publishing activity, *text publishing* might be characterized as the creation and refinement of documents that consist exclusively of text. Work in this category can be accomplished satisfactorily with any of several word-processing programs and almost any modern personal computer.

Graphics publishing is more demanding, in that it is centered around pictorial elements: illustrations, graphs, and the like (cf. Chap. 11). Successful and convenient work in this area presupposes the availability of a somewhat more elaborate computer system, including output devices (screen, printer) with high-resolution graphics capability, multiple software packages, and perhaps a scanner for the electronic capture of printed or hand-drawn images (cf. Chap. 8).

The third category, *page makeup*, encompasses the preparation of complex types of camera-ready copy. This is usually preceded by extensive work with specialized software for the creation of text, illustrations, and such miscellaneous graphic elements as formulas and equations. Only later are the separate pieces combined, formatted, scaled, and positioned with the aid of high-level page-layout software (cf. Chap. 10). In effect, the computer screen then becomes a substitute for a traditional printer's makeup table. Many have come to equate this type of work with one particular software house—Aldus Corporation—due to that company's pioneering development of the program PAGEMAKER[27] (cf. Sec. 10.2). PAGEMAKER was first introduced in 1985 for the MACINTOSH line of computers, but versions have since become available for IBM-compatibles as well. PAGEMAKER is credited with being the first page-layout program that was both powerful and at the same time relatively user-friendly, though it now competes with such formidable rivals as QUARKXPRESS (Quark, Inc.), READY,SET,GO! (Lettraset), and VENTURA PUBLISHER (Xerox), among others.

Electronic documents of all types must ultimately be subjected to expression in hardcopy form through the medium of a suitable high-resolution printer (cf. Chap. 6), generally a laser printer (cf. Sections 7.2 and 7.3). Alter-

---

[27] Advertisements for the program have sought to capitalize on this singular reputation by declaring that "PAGEMAKER *is* desktop publishing".

natively, diskettes or other storage devices[28] containing document *files* may be submitted to a commercial printing establishment for reproduction via an electronic photo-offset device (cf. Sec. 7.5).

Contrary to some predictions, the impact of desktop publishing techniques has not been limited to individuals and small companies: they have also become important in the professional production of books and especially periodicals by established and reputable publishing houses. Traditionally the various tasks associated with a publishing house, including all aspects of manuscript processing, were performed by in-house specialists, and the author's role was limited to establishing the *content* of a given work. In addition to the author(s), a conventional publication team might consist of editors, managers, production supervisors, layout personnel, typesetters, photographers, artists, printers, bookbinders, and others, all working together in close cooperation to bring a particular publication to fruition. Today it is not unusual for authors to participate more fully in the process, permitting the professionals to focus much of their attention on offering critical advice regarding the preparation of camera-ready manuscripts.

There remains one serious obstacle to full realization of the apparent potential in this approach, however, and it has been accorded too little attention in the media. Superficial use of the current generation of publication software is deceptively straightforward, and novices are easily misled into thinking themselves fully qualified to achieve a level of perfection worthy of a professional. Unfortunately, there is often a wide gulf between what is theoretically possible with a set of "DTP tools" and the product turned out by a well-intentioned amateur. Marvelous results are indeed achievable from a technical standpoint, but access to the most perfect typesetting system (cf. Part II) represents no guarantee that it will be handled with sensitivity. The euphoric attitude that once prevailed—especially in the years 1986/87—has subsequently given way to more down-to-earth realism. Part of this change can be traced to the fact that early sales projections for "DTP systems" proved unrealistic, and producers of both hardware and software are now faced with more sober projections. This has led them to redirect their attention to more limited goals: providing systems that make it easier, faster, and less costly

---

[28] Diskettes (Sec. 5.2) have the advantage of being inexpensive, but their relatively limited capacity (< 2 MBytes) and susceptibility to failure make them less than ideal. Removable hard disks (cf. Sec. 5.3), for example, represent one alternative that is more resistant to shock; some are reportedly able to withstand easily a fall of several feet onto concrete!

to prepare relatively attractive, largely informal, but nonetheless effective printed matter. As a result, it is no longer fashionable to make somber predictions of the impending demise of the world's leading book publishers.

One other important development deserves at least cursory mention: the rapid advance of powerful networking systems (cf. Sec. 4.5), which in some settings have transformed desktop publishing into a team activity conducted within the context of a broader organization. Some types of publishing have thus moved away from the "desktop" and into a "workgroup". The resulting hybrid known as *workgroup publishing*[29] is particularly well-adapted to editorial offices charged with producing newspapers and periodicals. Various aspects of the editing process—including text acquisition and revision, graphic design, preparation of advertising copy, and page makeup—may still be parceled out among experts, but the parts remain fully integrated within a single "editorial network" consisting of multiple interconnected workstations.

1.4 Applications of Desktop Publishing

We have stressed repeatedly that anyone with access to a modern personal computer and a few key accessories is potentially capable of producing printed copy analogous to that from a professional print shop with hundreds of thousands of dollars worth of equipment—except for the problem of displaying equivalent *design* skills. Desktop publishing is in fact ideally suited to situations in which economy and timeliness are paramount, with optimum layout perhaps playing a secondary role.

DTP technology can assist the individual "communicating" scientist or engineer at nearly every stage of his or her career, from the writing of reports and theses through the later arduous but perhaps more important tasks of preparing grant and patent applications, research papers, and even books.[30]

---

[29] For a recent discussion of this subject see the editorial by Sandra Rosenzweig in the December 1991 issue of *Publish*.
[30] For an early example of a technical book the final copy for which was prepared almost exclusively by its authors see the ca. 500 page *Art of Scientific Writing* (Ebel, Bliefert, and Russey, 1987).

Business organizations have found a great many creative uses for DTP, including the preparation of attractive and informative

- advertisements, prospectuses, and brochures
- data sheets
- descriptions of products and services
- operating instructions
- catalogues
- price lists
- press releases
- forms
- employee handbooks
- visitor orientation plans
- corporate profiles
- confidential reports
- invitations
- internal memos
- annual reports
- in-house newsletters
- summaries of employee benefit packages
- posters and placards

One very important application of DTP in the business setting is the preparation of handbooks, catalogues, and other documents subject to frequent updating. Material of this nature can easily be manipulated in such a way as to provide a product that looks truly professional even as it remains subject to extensive last-minute modification.[31] Probably the most important prerequisite for success in this application is the ability to transfer data smoothly and reliably from a company's central computers into specialized sub-systems dedicated to work with text and graphics.

Particularly in a corporate setting, speed and efficiency are critical parameters, and it is precisely here that desktop publishing shines: by sharply curtailing the necessity for sending copy back and forth between authors, management, typesetters, graphic designers, and clerical workers at several stages of the editing process. A good example is provided by the publishing

---

[31] An interesting example is the mail-order catalogue for the software house "MacWarehouse", an attractive, colorful, informative publication of some 200 pages appearing at frequent intervals and prepared exclusively by DTP techniques.

house responsible for this particular book—which has been very receptive to our utilizing the kinds of techniques we advocate (cf. Appendix C). Our publisher, VCH, installed a pilot in-house DTP center during the spring of 1988, linking it from the outset with the company's central publications database of bibliographic and descriptive information on all new and forthcoming books and journals. Known as "BISy" ("Book Information System"), the system required an initial investment of over $10 000, but it was found that it paid for itself within less than a year!

Publishers in general have come to recognize that DTP techniques can also make a positive contribution to their primary activity: publishing. Many newspapers are now prepared by DTP workgroups, and the same is true for a broad array of magazines, technical journals, and other types of serial publications. Even the release of high-quality, attractive books based on camera-ready copy is no longer a rarity. Indeed, one might well argue that the rigid distinction between DTP and "professional" publishing methods is becoming increasingly anachronistic, and perhaps counterproductive.

1.5 Advantages and Costs Associated with Desktop Publishing

Several important benefits can be expected to arise from the decision to substitute DTP techniques for conventional typesetting. These include:

- Direct access to and control over all aspects of the production process, often resulting in products containing fewer errors

The personnel assigned to a DTP center assume full responsibility for every phase of production, including layout, typesetting, graphics preparation, page makeup, and printing. Publication content can be kept exceptionally current, because changes are possible literally at the last minute. Every document remains totally accessible to its author(s) until the final stages (i.e., beyond the point of typesetting), which greatly reduces the risk that errors will be overlooked. Desktop publishing also confers upon all the parties involved a welcome sense of independence.

● Significant savings in time

The awkward and time-consuming step of sending copy and proofs back and forth between typesetter and author is eliminated completely. Once text has been input into the computer the author/client can begin immediately to read his or her own proof, make corrections, and adjust the graphics or even the layout, all on the basis of images displayed on a computer screen. Test prints can then be prepared with an adjacent laser printer, and these can be examined minutely for possible problems. This step is in fact one of the most important, especially if final production is to involve a more sophisticated printing device, since slight discrepancies must always be anticipated between a screen image and its printed counterpart (cf. Sec. 6.3).

Publishers report that a procedure like that outlined above can lead to time savings of as much as 40%, for example, in the preparation of a technical journal. The savings are likely to be less dramatic with documents requiring significant amounts of color printing, or if large numbers of high-quality half-tone images are involved, but even here DTP technology has taken dramatic strides in recent months. Still, fairness requires pointing out that complicated layouts for heavily illustrated, multicolored brochures can in some cases even today be achieved more rapidly or more satisfactorily by conventional means (in this context see the valuable critical observations of Peters, 1988, and the comments of Ken Draeger in the February 1991 issue of *Publish*).

● Low production costs

The in-house application of DTP technology often results in considerable savings relative to outside contract printing. A simple desktop-publishing workstation, including a computer, all the necessary software, a graphics-compatible laser printer, and even a scanner, can now be acquired (Spring, 1994) for under $6000. This refers to a facility fully equipped to carry out every stage of the process—from the creation of copy (both text and graphics), through typesetting and page makeup, to the production of near-typeset quality printed pages. It is also worth noting that personal computers and laser printers rarely experience costly down-time, and when repair work does become necessary the cost is usually modest. Output quality can be enhanced materially by the acquisition of a computer-driven laser typesetter (cf. Sec. 7.5), but devices in this category remain quite expensive (ca. $20 000–$200 000), and they are somewhat difficult to operate and maintain. The fact that initial DTP investment costs are relatively low means that a small company or institution can often afford to conduct an experiment in publishing its own periodicals, brochures, or even books without incurring undue risks. Substantial savings can sometimes be realized even if final production of copy prepared in-house is contracted out to a commercial printer, who uses the DTP-generated data files as input for a phototypesetter.

● Modest outlays for personnel training

A major advantage (often stressed heavily by the distributors of DTP hardware and software) is that effective use of the new technology is possible with little or no professional training.[32] It is undeniably true that an amateur can produce reasonable printed copy much more readily with a personal computer and a page-layout program like PAGEMAKER than with a professional typesetting system. Part of the difference lies in the fact that a DTP system provides the opportunity to examine preliminary versions of a work directly on the screen (WYSIWYG; cf. Sec. 6.3), and then modify them as desired. Nevertheless, in order to achieve results that would win the approval of a professional, the average beginner will find it necessary at least to read a few basic books on typography and layout, and perhaps attend a well-conceived course on the subject. Special training courses in page layout for DTP, typically a few days in length, are now offered on a regular basis by a number of service organizations. A staff person newly assigned the responsibility for preparing company fliers, catalogues, or similar documents would be well advised to enroll immediately in such a course, and also seek close contact with an experienced professional graphic designer willing to critique the initial efforts. Often the best policy is to entrust at least part of the early design work to a qualified professional from the outset.

Individual scientists or engineers interested in preparing camera-ready copy of their work should similarly seek professional guidance from someone in the production office of a publishing house. This contact should occur

---

[32] The other side of the coin is the subject of the following harsh words from an especially sensitive critic (published in *CT Magazin für Computertechnik* **1987**, *9*, 78): "… In the wake of the apparently unending trend toward computerization, MACINTOSH equipment from Apple has played a unique role in reducing typographic variety to its most primitive elements: always the same decorative lines and vignettes, monotonous textures and patterns, a casserole of fonts and type sizes in every conceivable and even impossible form (bold, italic) and embellishment. Anyone acquainted with the frightening results of tests conducted with school children will look with skepticism toward the typographic future."

before work begins, and one should certainly not be hesitant to ask additional questions later as the project matures. It is our hope that careful reading of the next two chapters of this book will help many novices to avoid at least some of the most flagrant errors.

Practicing typesetters and graphic designers have the great advantage of years of experience, during which they will have learned many of the formal traditions (and tricks) of their trade. One of these is the unfamiliar habit of examining every printed page from an artistic and aesthetic point of view. Hardware and software firms may continue to contend that little or no professional training is required for electronic editing, but—fortunately—the opposing viewpoint also has its advocates, and even publishing houses have sometimes been forced to admit that their in-house DTP needs are best served by specially trained personnel.[33]

● Image enhancement

The ability to project in printed form a polished, professional image was once beyond the means of those responsible for minor publications, such as newsletters, technical handouts, or product descriptions.[34] Here DTP technology represents a truly major breakthrough—assuming the availability once again of appropriate design skills.

● Efficiency

Advertisements conceived by Aldus, the developer of PAGEMAKER software, stress the fact that the cost advantages of DTP come not only through consolidation in the production process, but also with reduction in the number of steps involved. Conventional preparation of a single page of a professional journal is said to be more than twice as expensive as reproducing a similar page prepared by the author. American book publishers also report savings through DTP on the order of 50%, although critics like Peters (1988) have argued that such analyses tend to be oversimplified and misleading. An important prerequisite to realizing maximum savings in a business setting is

---

[33] Interestingly, many of the advertisements used by DTP firms themselves are prepared with the aid of professional typesetters and graphic designers (cf. Peters, 1988).

[34] The desire for "professional-looking" printed materials is largely responsible for the dramatic proliferation of small "copy shops" and specialty printers equipped to provide "laser-quality" output quickly and inexpensively, and in any desired quantity, based on a customer's own originals.

the establishment of full compatibility between a company's DTP stations and other in-house data-processing equipment.

1.6 Desktop Presentation

One other application of computers to text and graphics, *desktop presentation*, actually differs very little from desktop publishing in either the goal or the requisite hardware (cf. Cole and Odenwald, 1989). The object in this case is to exploit computer-based graphic-design techniques in the assembly of text and numerical data in a form suitable for display on a video screen or by optical projection (from 35-mm slides or overhead transparencies), usually to supplement a lecture or similar oral presentation. Alternatively, such information might be made available for distribution in the form of handouts. Special programs described as *presentation software* have been developed to meet this need as an alternative to ordinary word processors and conventional graphics and layout programs. The challenge has been to make it as easy as possible to produce visual enhancement for an oral presentation— even a dramatic "multimedia show", complete with moving (video-screen) pictures, sound effects, and professional-looking static images.

Software of this type can be used to great advantage by industrial and academic scientific personnel in conjunction not only with formal lectures but also more routine oral reports. Success as always presupposes that the user possesses a rudimentary feeling for effective design, and that the underlying message is well organized, but conscientious application of presentation software can also *contribute* to better organization. A typical program in this category combines

- facilities for the hierarchical structuring of a series of topics, organized in outline form
- graphics tools like those in dedicated "draw" and "paint" programs (cf. Chap. 11)
- a complete set of basic word-processing tools
- integral program functions for subjecting numerical data to useful transformations (cf. also Sec. 12.4), leading to various familiar types of "business graphics"

● spreadsheet features that make it possible to illustrate sample computations

Several desktop-presentation packages are available, each with its own unique approach to the structuring and subsequent display of information. One of the first (whose release actually pre-dated the term "desktop presentation") was originally called THINKTANK, but now goes by the name MORE (from Symantec). MORE might also be characterized as an excellent example of *outlining software*, and it can be a valuable asset at the very earliest stages of collecting and structuring one's ideas and thoughts, as well as later when the time comes to prepare supplementary slides or transparencies. The user interface for MORE is illustrated in Figures 1–5 and 1–6. Programs of this nature can also contribute to the preparation of a conventional written document, consistent with the observations "… a scientific paper is primarily an exercise in organization" and "good organization is the key to good writing" (Day, 1983).[35]

Other presentation programs—available in both MACINTOSH and IBM-compatible versions—include CA-CRICKET PRESENTS (Computer Associates) and PERSUASION (Aldus), which offers an elaborate "slide-show" function. Not surprisingly, the availability of such software has also led to increased interest in *slide makers*, relatively expensive devices for efficiently transforming multicolor computer graphics into high-resolution projection slides.

Additional MACINTOSH programs useful in the desktop presentation context include HYPERCARD (discussed in detail in Sec. 12.5) as well as ADOBE PREMIER (Adobe systems) and MOVIEWORKS (Interactive Solutions), both of which facilitate the preparation of computer-based motion pictures. Programs of the latter type make it possible—perhaps with the help of a standard graphics package and/or HYPERCARD—to assemble complex animated screen presentations (demonstrations, for example).

1.7 Future Prospects

It seems safe to assume that the future holds many exciting possibilities for the world of desktop publishing, as computers themselves become faster (cf.

---

[35] The developers of MORE speak of it as software for "idea processing and presentation".

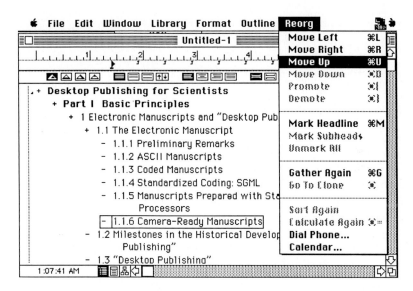

Fig. 1–5. An illustration of some of the techniques available for structuring a text document with the program MORE.

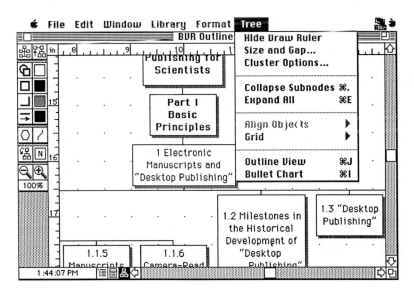

Fig. 1–6. Alternative graphic representation of one segment of the MORE document illustrated in Fig. 1–5.

Chap. 4), screen resolution increases (cf. Sections 6.2 and 6.3), and memory capacity continues to expand. The latter consideration is particularly important with respect to sophisticated graphics, especially graphic images based on three dimensions. The potential inherent in small systems like those that are the subject of this book can be expected increasingly to approach that of professional typesetting equipment.

At the same time, manufacturers of commercial typesetting equipment will continue to incorporate into their own products various features that have proven their worth in the course of the "DTP revolution". Several aspects of software developed for the DTP market have already stimulated the imaginations of professional printers and typesetters, foreshadowing interesting developments in the field of commercial publishing. These include novel approaches to the "text wrap" problem (arranging text around graphic elements), the potential for stretching or otherwise deforming type in every conceivable way, a virtually unlimited range of type sizes (e.g., from 3 to 500 points, including fractional values; cf. Chap. 2), and full-screen page previews, complete with graphic elements.

In the case of peripheral equipment, laser printers (cf. Chap. 7) with resolution significantly higher than the current standard of 300 dpi will certainly become commonplace within a very few years. Scanners (cf. Chap. 8) will also be capable in the future of capturing images at increasingly high resolution, and scanning speeds will continue to increase—as will the proficiency of software for interpreting scanned text.

At the same time, growing numbers of books and periodicals are likely to be produced on the basis of electronic manuscripts prepared completely by their authors, and the resulting documents will be almost impossible to distinguish from publications prepared in the traditional way, particularly as wider use is made of DTP-compatible laser typesetters (cf. Sec. 7.5).

The role of color in desktop publishing, a subject of great media attention, is really still in its infancy, although high-quality color monitors, color scanners (cf. Chap. 8), and, more recently, PostScript-compatible color laser printers are becoming increasingly available and affordable. Experts contend that the results achieved so far remain inferior to what can be accomplished professionally, but the gap is certain to shrink with further developments in both hardware and software, and vivid color can be expected to become a standard element in desktop presentation materials.

2 Typography and Layout

2.1 Introduction

In this chapter we focus our attention on various aspects of *typography*. The word itself is derived from the Greek τυποσ *(typos)*, meaning "form" or "pattern", and γραφειν *(graphein)*, "to write". In the broadest sense it encompasses every aspect of the conscious distribution of text over a predefined surface—in short (at least for our purposes), typography is "*the art, or skill, of designing communication by means of the printed word*" (McLean, 1980). Typography at its most basic level is concerned with the design characteristics of individual letters, but for most desktop publishers typography begins with the act of selecting a set of typefaces and type sizes, setting dimensions and boundaries for blocks of text, and establishing layout standards applicable not only to the body of a document but also to the *front matter* (preface, table of contents, etc.) and other subsidiary elements (figure captions, tables, running heads, etc.). Typography usually culminates in the assembly of individual pages, including careful placement of tables, captions, footnotes, and the like. Typographic considerations are also brought to bear on relationships between the text itself and blank spaces, line drawings, photographs, ornamental embellishments, structural lines, etc. The broadest definition of typography extends to the choice of paper and binding—even the *concept* associated with a printed work (its overall character, as reflected, for example, in promotional matter and advertisements).

More often the word "typography" is applied in a narrower sense, however, in which it is restricted to the formal treatment of text elements in a typeset document. Professionals sometimes make a distinction between "microtypography" and "macrotypography". *Microtypography* (or *detail typography*) refers to letters, letter spacing, word spacing, line spacing, and similar matters, whereas *macrotypography (design)* is concerned with more general aspects of formatting and layout, including overall dimensions, placement of text on the page, graphic elements, and the stylistic treatment of figure captions and legends.

This chapter addresses the key elements of *book typography* as viewed from the perspective of the aspiring desktop publisher: typographic considerations fundamental to the effective design and development of a monograph, perhaps, or of a more modest work such as a report or a journal article. Our attention will be directed mainly toward text elements—especially text of a scientific and technical nature—at the expense of more peripheral subjects like bookbinding or the design and preparation of promotional material.[1]

The single most important goal of any typographic effort should be enhancing *communication* by providing appropriate and attractive surroundings for the message that constitutes the heart of a document, thereby assuring optimum readability.[2] For this reason we begin by considering briefly the reading process itself.[3]

---

[1] For an overview of typography with an emphasis on traditional methods see Craig (1992). An excellent source of further information, especially for the desktop publisher—presented in an elegant way—is the book *Graphic Design for the Electronic Age: The Manual for Traditional and Desktop Publishing* (White, 1988). Another almost indispensable guide is *The Thames and Hudson Manual of Typography* (McLean, 1992). A marvelous collection of very practical advice and suggestions, interspersed with wry philosophical comments, is *The Elements of Typographic Style* (Bringhurst, 1992), whose poet/typographer author frankly acknowledges an indebtedness to Strunk and White's famous little book on writing, *The Elements of Style* (1979). Bringhurst's book is dedicated to "writers & editors, type designers, typographers, printers & publishers, shepherding words and books on their lethal and innocent ways." One of the most comprehensive English-language sources of general information on all aspects of desktop publishing—including typography—is the book by Kleper (1990), *Desktop Publishing and Typesetting*. Also worth mentioning is Brown (1989), as well as the book by Baumann and Klein (1989), in part because the latter constitutes in itself an excellent specimen of attractive and professional desktop publication. For an overview of all facets of the traditional approach to book publishing, including editing, see Lee (1980).

[2] Some make a distinction between the two closely related terms "legibility" and "readability", but we follow what appears to be the current trend of treating them as synonymous (cf. Kleper, 1990).

[3] Most of the subsequent remarks on the subject of reading are derived from the useful little booklet *Das Detail in der Typografie* (Hochuli, 1987). Relevant information can also be found, for example, in McLean (1980), Lieberman (1967), White (1988), Korger (1977), Rüegg (1972), and Gerstner (1985), as well as in articles from the journal *Visible Language* (formerly the *Journal of Typographical Research*). For a particularly scholarly account of the subject see Spencer (1969).

The act of reading begins with a succession of abrupt, rapid motions *(saccadic jumps)* of the eye as its focal point traverses a printed page. These motions are interrupted by brief periods of fixation (ca. 0.2–0.4 s long) during which bits of visual information (5–10 letters; i.e., one or two words) are collected and transmitted to the brain. The beginnings and ends of lines serve as important signposts to guide the eye in the course of its journey. The more skilled the reader, the shorter are the fixation periods and the more frequent the saccadic jumps.

Written text can be absorbed in at least two different ways. In *true reading* individual letters are perceived directly, after which they are grouped, interpreted, and incorporated into thoughts, whereas in *pictorial reading* thought connections are established on the basis of more complex visual patterns. Here the printed word functions more like a "picture", permitting shapes and forms to be correlated rapidly and unconsciously with *other* word pictures already stored in the reader's "visual memory" as a result of previous reading experiences.

It is the goal of *legibility research* to establish objective criteria for evaluating the legibility of a printed work, with considerable attention devoted to the study of eye movement. The rate at which a given piece of text can be read and interpreted has been found to be closely related to many factors, including line length, type size and style, and visual contrast between letters and their background. It has also been shown that subjective judgments about legibility tend to agree quite well with more objective data. One elementary finding—easily verified—is that TEXT SET ENTIRELY IN CAPITAL LETTERS IS RELATIVELY DIFFICULT TO READ, in part because all letters are the same height, leading to rather undifferentiated "word pictures". Similarly, large blocks of text seem to be easiest to read when set in a typeface that includes *serifs* (tiny embellishments at the ends of the letters, as in the *Times* face used generally throughout this book). A *sans serif* face, like the one (*Helvetica*) employed for this particular sentence, offers fewer visual cues, making the corresponding "word pictures" less distinctive (cf. Sec. 2.3). Serifs are also thought to assist the eye in holding to a smooth horizontal course.

In the sections that follow we explore many other concepts and principles that contribute to both legibility and aesthetic satisfaction, in the process introducing a few of the more important technical terms associated with the printer's art. Rules we cite are based on long-standing practice, and are firmly embedded in typesetting tradition. We will obviously be able to do little more than scratch the surface of what is a very extensive landscape, and we certainly harbor no illusions about transforming our readers into professional typesetters. Nevertheless, conscientious consideration of these paragraphs and consistent adherence to the suggested rules should prevent the aspiring desktop publisher from making some of the most common mistakes. While creating an acceptable piece of camera-ready copy certainly does not require mastery of all the subtleties of typography, or a firm grasp of the technical distinctions between, for example, the "letterpress" and "offset" printing processes, it is important to be familiar with the basic principles.

2.2 Fundamentals of Page Layout

In typography, as in any art, it is often the details that count most. The overall effectiveness of a printed work is determined by a host of interrelated factors that combine in mysterious ways to confer upon each publication its own unique character.[4] Much more is involved in the success of a document than simply the intellectual message embodied in its words. One key element is the nature of the *type* in which the work is set (including its size and spacing),[5] but equally important are *word spacing*, *line spacing*, *line length*, and the *arrangement* of text blocks and illustrations, all of which in turn contribute to the ultimate *size* of the overall work.

Once decisions have been made about basic parameters such as these, the most important rule to bear constantly in mind in the attempt to create a technically "clean" piece of work is: be consistent. All the characteristic features of a particular layout should be strictly maintained from beginning to end. Consistency of style is what endows a book with a uniform "feel". This demand for consistency applies quite generally, affecting

- typeface(s)
- type style(s)

---

[4] "Typography is to literature as musical performance is to composition: an essential act of interpretation, full of endless opportunities for insight or obtuseness." (Bringhurst, 1992)
[5] "… type can destroy a text just as easily as it can lead to its perfection." (Luidl, 1988)

- line spacing
- justification
- margins
- column layout
- running heads
- tables (general structure and placement)
- illustrations (position, form—lettering, line weight, etc.—and captioning)
- overall page design

Type size, line length (or line *measure*), line spacing, and line count are the variables most obviously associated with the text itself, and these will be discussed in some depth later (Sections 2.4 and 2.5), but before such parameters can be established a decision must first be reached about the appropriate *page size*. Page-size selection reflects a balance between many considerations, and different publishing houses tend to prefer different formats. In the absence of external guidance, the best course for a desktop publisher to follow is to examine a number of works roughly comparable to the one envisioned, and then choose a suitable model (cf. also Sec. 2.13).

Once a page size has been adopted the next thing to consider is the optimal *live-matter area*[6]—the amount of the page surface that will be devoted to text—and the way the remaining space will be distributed between the four margins. It is always a mistake to try to put as much text as possible on a single page. The goal instead should be to create a document that is comfortably *readable*. Blank space around the edges gives "life" to a text, permitting it to "breathe". The impression conveyed by a piece of text can vary dramatically depending upon the amount of page space it occupies: "compressed" if margins are too narrow, "lost" if they are too wide. One rule of thumb suggests that a page containing both text and illustrations should reflect a ratio of printed to unprinted surface of at most 4 : 1.

Figure 2–1 offers a fairly traditional guide to positioning text on a printed page. Note that the inner (or *gutter*) margin is the narrowest, that at the top *(head)* is a bit wider, and those at the outside *(fore-edge)* and bottom *(foot)* are wider still. A good approximate relationship for these four margins is 2 : 3 : 4 : 5 (for more information see, for example, Bringhurst, 1992, White, 1988, or Luidl, 1984). In the case of letter-size paper (8.5″ × 11″) the nar-

---

[6] In the case of a book this is defined in such a way that it does *not* include space assigned to page numbering, although the latter is counted as part of the *type page*.

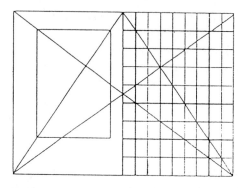

Fig. 2–1. A common geometric approach to establishing the margins for a printed page.

rowest margin should be at least 0.5″ wide. This permits the reader to hold the pages comfortably without obscuring any of the text, and at the same time facilitates stapling or binding.

Another important consideration from an overall design standpoint is whether or not to employ a *multiple column* text format. The main advantage of breaking text into columns is that it prevents the lines from becoming too long, since long lines are inherently difficult to read. If one does elect to divide the text into columns it is important to ensure that there is an appropriate amount of space *between* the columns, taking into account both structural clarity and aesthetic balance. In the case of a work that consists almost exclusively of text it is sometimes desirable to emphasize column separation by the introduction of thin vertical dividing lines.

Most DTP layout programs offer what are known as "master pages" to simplify the task of maintaining layout uniformity throughout a multipage document (cf. also Sec. 10.1). Elements or parameters that are applicable to every page—column dimensions, pagination, margins, structural lines, etc.— need only be specified once on the (left and right) master pages, after which corresponding features are incorporated automatically into successive pages as they are created. In a similar way, many programs permit one to establish at the outset typographic characteristics applicable to various kinds of *paragraphs*—line spacings, type sizes, indentations, etc., for normal paragraphs, freestanding equations, figure captions, section headings, and the like. These are defined in what is known as a *style sheet.*[7] Subsequently applying

---

[7]   Styles, once defined, are always subject to later modification. Thus, a document structured in terms of styles can be easily reformatted at any time. For example,

one of the predefined *styles* to any specific paragraph causes that paragraph to acquire immediately all the appropriate typographic characteristics.

The information that a document is intended to convey, especially if it is a scientific work, is usually organized *hierarchically*, and it is important that the typography reflect and reinforce such organization. Care in this regard is particularly appreciated by the casual reader, who at first may wish only to acquire an overview of the contents, but organizational clarity also helps focus the attention of a reader seriously intent upon mastering the subject. Hierarchical document structure, like an outline, acknowledges the fact that not every text element is of equal importance. Titles and section headings deserve more recognition than footnotes, for example, and in a technical discourse certain words ("keywords") are often more important than others. Such distinctions deserve to be supported by a book's typography. Typographers have developed several strategies—some quite subtle—for providing the reader with insight into a document's structure. *Boldface type* is an example of a not-so-subtle typographic signal: by its very nature it causes words to stand out from the surrounding text. Boldface type must therefore be employed sparingly and judiciously—exclusively in headings, for example—because overuse drastically reduces its effectiveness. We return to this subject in Section 2.3.

An orderly and consistent arrangement of text elements is one of the most important hallmarks of a good layout. A systematic layout assists the reader in an unobtrusive way in distinguishing and identifying individual units of information within a document (paragraphs, lengthy quotations, illustrations, captions), and then in categorizing and intuitively assigning a relative degree of importance to each. The goal of clarity must be approached with caution, however. Pages with too many contrasting visual elements may instead foster confusion, leaving the reader's eye incapable of distinguishing rapidly between the important and the trivial. Assessing the effectiveness of a layout requires one to look at the pages from a distance, evaluating each as one would a picture. Components should then be rearranged as necessary in an appropriately balanced way—bearing in mind that readability must also remain an important criterion. As suggested above (and in several of the

---

the typeface assigned to second-order headings can be changed uniformly and instantaneously throughout a document by issuing one simple command. The principle is the same as that underlying the "generically coded manuscript" discussed in Section 1.1.4.

sections that follow), one of the most critical factors in achieving a satisfactory page layout is the successful integration of blank space.

An extra measure of structure is sometimes introduced into an especially complicated work in the form of brief *marginal notes* set outside the standard type zone of the page, offering the reader a rudimentary outline of the content. Typesetting such notes can be awkward and time-consuming, but the reader interested in quickly finding a particular piece of information will welcome the extra consideration.

Finally, mention should be made of the possibility of occasionally setting

a specific text element against a dot-screen (gray) background

for emphasis, or even incorporating a few blocks of

white lettering set against a black background.

Effects like these can be very striking, but they must be introduced with care and in a systematic way to prevent pages from acquiring a "busy", cluttered look.

2.3 The Type Itself

The basic typographic units of the English language are the 26 letters of the alphabet, which can be expressed in two typographic variations: "capital" and "small".[8] A letter is really nothing more than a universally agreed upon pictorial symbol for evoking graphically one of the sounds in a spoken language. A given *typeface* constitutes one particular set of such symbols (cf. Fig. 2–2). Equivalent characters from different typefaces can be distinguished on the basis of their

---

[8] Capital and small letters are also referred to as "uppercase" and "lowercase", respectively, recalling an earlier time when type was set by hand, and the metallic type elements were stored in compartmented wooden cases—with the capitals on top. An even older pair of terms is "majuscule" and "minuscule". A unique and lavishly illustrated source of information on letters and linguistic symbols from throughout the world—present and past—is the book by Faulmann (1990), a reprint of an original 1880 edition. Jean (1992) has provided an interesting and very beautifully illustrated little paperback (published originally in French) dealing with the historical development of writing and printing.

- general shape or *family* (e.g., Times, Palatino, Bookman)
- *style* (roman, italic, etc.)
- *line weight* (e.g., fine, normal, bold)
- *height* (a subject discussed in some detail in Sec. 2.4)
- *breadth* (narrow, normal, wide, etc.), as with "Helvetica normal" and "Helvetica narrow"

Typefaces can also be classified more broadly in terms of whether or not the corresponding letters feature *serifs*—extra finishing strokes that assist the eye in making its way through a text passage, thereby increasing legibility.

Times roman
abcdefghijklmnopqrstuvwxyz!""'$%&/()=?
ABCDEFGHIJKLMNOPQRSTUVWXYZ1234567890

Times italic
abcdefghijklmnopqrstuvwxyz!""'$%&/()=?
ABCDEFGHIJKLMNOPQRSTUVWXYZ1234567890

Times bold
abcdefghijklmnopqrstuvwxyz!""'$%&/()=?
ABCDEFGHIJKLMNOPQRSTUVWXYZ1234567890

Times bold italic
abcdefghijklmnopqrstuvwxyz!""'$%&/()=?
ABCDEFGHIJKLMNOPQRSTUVWXYZ1234567890

Helvetica roman
abcdefghijklmnopqrstuvwxyz!""'$%&/()=?
ABCDEFGHIJKLMNOPQRSTUVWXYZ1234567890

Helvetica italic
abcdefghijklmnopqrstuvwxyz!""'$%&/()=?
ABCDEFGHIJKLMNOPQRSTUVWXYZ1234567890

Helvetica bold
abcdefghijklmnopqrstuvwxyz!""'$%&/()=?
ABCDEFGHIJKLMNOPQRSTUVWXYZ1234567890

Helvetica bold italic
abcdefghijklmnopqrstuvwxyz!""'$%&/()=?
ABCDEFGHIJKLMNOPQRSTUVWXYZ1234567890

Fig. 2–2. Two of the typefaces used most frequently by desktop publishers, *Times* and *Helvetica*, each depicted in several different type styles.

Letters in certain typefaces lack this kind of embellishment (*sans serif* type; cf. Sec. 2.1). The distinction is highlighted in Fig. 2–3.[9] It is often asserted that readers find it easier to read large amounts of text set in type that includes serifs, though not all authorities agree (cf. White, 1988).

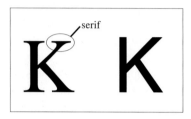

Fig. 2–3. An example of a capital letter with and without serifs.

Publishers sometimes refer to the typefaces most commonly exploited in their trade as *book faces*. A book face is distinguished by the fact that its letters are exceptionally legible, but also relatively unobtrusive (what matters most in a typeset paragraph is *content*, not form!). Generally speaking, a good book face should be clearly readable at a distance of about 14 inches. A much larger typeface category encompasses the *special faces*, those most often associated with applications purposefully designed to attract attention (advertisements, business cards), or where recognition at a distance is critical (cf. Fig. 2–4).

Typefaces also lend themselves to broad categorization on the basis of *size* (for an explanation of the "point" as a unit of measure see Sec. 2.4):

- small type (6–8 points)
- text type (9–12 points)
- display type (14 points or larger)

Type in the first category is useful when the need arises to compress a large amount of information into a limited space, as in a dictionary or a telephone

---

[9] Depending upon the nature of the serifs, further classification is possible into "small serif", "semi-serif", "split-half-serif", and many other categories (for more details see Lieberman, 1967). The delightful booklet by Spiekermann (1986), described in its foreword as "the autobiography of an impassioned typographer", contains the observation: "Serifs are in a sense flowers on the balcony of the type; by contrast, the more perfect and timeless sans serif types often possess little more charm than a clean but boring street in the suburbs."

Charlemagne
ABCDEFGHIJKLMNOPQRSTUVWXYZ1234567890

Herculanum
ABCDEFGHIJKLMNOPQRSTUVWXYZ1234567890

Duc de Berry
abcdefghijklmnopqrstuvwxyz!""$%&/()=?
ABCDEFGHIJKLMNOPQRSTUVWXYZ1234567890

Kuenstler Script Medium
abcdefghijklmnopqrstuvwxyz!""$%&/()=?
ABCDEFGHIJKLMNOPQRSTUVWXYZ1234567890

Zapf Chancery
abcdefghijklmnopqrstuvwxyz!""$%&/()=?
ABCDEFGHIJKLMNOPQRSTUVWXYZ1234567890

Birch
abcdefghijklmnopqrstuvwxyz!""$%&/()?
ABCDEFGHIJKLMNOPQRSTUVWXYZ1234567890

Blackoak
abcdefghijklmnopqrstuvw
ABCDEFGHIJKLMNOPQR

Madrone
abcdefghijklmnopqrstuvw
ABCDEFGHIJKLMNOPQR

Fig. 2–4. Examples of several distinctive "special" *(display, decorative)* typefaces.

directory, but it can also be called upon for footnotes, marginal notes, or running heads (cf. Sec. 2.9). Excessive reliance on type that is too small is a disservice to the reader, however, because small type is very tiring to read. At the other extreme, unusually large type is inappropriate for body text because it adds too much to the page count—and the cost—of a printed work. As a general rule, running text should never be set larger than about 14 points or smaller than 8 points. A good compromise in most situations is 10-point type.[10]

---

[10] For a more extensive discussion of type-size considerations, with emphasis on the role played by the probable reader, see Lee (1980), p. 91.

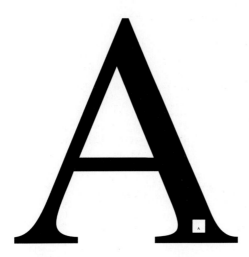

Fig. 2–5. An example of two extreme type sizes: 3 points and 256 points.

In sharp contrast to typewriters, the computer-based printing devices now available to most authors offer almost unlimited flexibility with respect to type size. Many even support fractional sizes, along with the potential for proportional shrinking and enlargement of both text and graphic elements (cf. Sec. 7.3). On the other hand, editorial *software*—word processors and layout programs—may impose type-size limitations of their own, such as a restriction to integral sizes and perhaps a minimum size of 2–4 points with a maximum of 127 or 256 points (cf. Fig. 2–5).

The operating system of a modern personal computer usually includes provisions for several different typefaces (or *fonts*),[11] some of which are optimized specifically for relatively low-resolution dot-matrix printers. With certain operating systems (notably the MACINTOSH system, Sec. 4.4) it is a very straightforward matter to increase the number of available "types of type". Activation *(installation)* of a new font may require the use of appropriate supplementary software (as illustrated in Fig. 2–6), but once installed that font becomes permanently and instantly accessible to virtually every text and graphics application.

---

[11] Today the terms "typeface" and "font" tend to be used interchangeably. Originally, *typeface* referred to the *design* of a set of characters, whereas a *font* was a physical set of metallic letters cast in some particular face. Modern "electronic type" is created on demand—and in almost infinite variety—by making appropriate adjustments to complex mathematical descriptions of the desired characters, virtually eliminating the grounds for distinguishing between "concept" and "product".

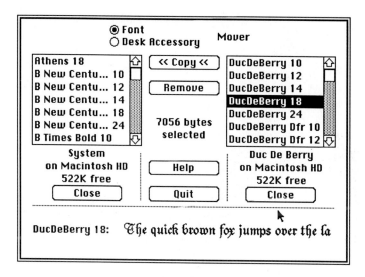

Fig. 2–6. Dialog box from the program FONT/DA MOVER, used for installing fonts in the MACINTOSH operating system (versions prior to System 7).

Flexibility with respect to type size and design is certainly a major step in the direction of "professional" typesetting, but it must also be emphasized that fonts accessible to desktop publishers rarely reflect the refinement or adaptability of analogous typefaces created for sophisticated photo-offset typesetting systems. For example, in most DTP fonts the only available *ligatures*—special combinations of two (occasionally three) adjacent letters, such as "fl"—are the few permitted within the confines of a 256-character limit.[12] Similarly, fonts for a personal computer seldom[13] offer a choice between *ranging* or *lining numbers* (all of whose digits rest on the baseline and are of equal width) and *old style* (or *lowercase*) *numbers*, analogous to lower-case letters in that they include a few *ascenders* (strokes that extend above

---

[12] For example, the MACINTOSH operating system provides uniform access only to "æ", "Æ", "œ", "Œ", "fi" and "fl" (code numbers 190, 174, 207, 206, 222, and 223, respectively, corresponding to the key combinations option-', shift-option-', option-q, shift-option-q, shift-option-5, and shift-option-6).

[13] Early in 1992 Adobe Systems announced the release of several special personal-computer font sets containing both old-style numbers and true small capitals [described in *Form & Function* (the Adobe biennial font newsletter), No. 10, 1992].

the x-height; cf. Sec. 2.4) and *descenders* (strokes ending below the baseline), e.g.:

0 1 2 3 4 5 6 7 8 9

Professional typefaces also tend to vary somewhat as a function of size—reflecting non-proportional adjustments to certain stroke widths, for example, or subtle adaptations in letter spacings. With a personal computer and word-processing or layout software one is usually limited to rather primitive algorithms for generalized size-dependent spacing changes *(tracking)*, although many programs do provide support for manual (or even automatic) *kerning*: selective adjustment of the space separating particular pairs of symbols. This is important, because excess blank space otherwise associated with certain letter pairs (because of the way the corresponding shapes happen to match) can be quite unsightly, especially with the large type used in titles or headings. Two examples of the problem and the way kerning can alleviate it are illustrated in Fig. 2–7. Professional typesetting equipment makes extensive use of detailed sets of spacing instructions *(kerning tables)* adapted not only to a multitude of letter combinations but also to various sizes.

Figure 2–8 provides samples of several serif and sans serif typefaces available for a typical laser printer. Most are of the *proportional* variety, in which different letters are assigned different widths (e.g., "i" vs. "m"). The font called *Courier* is an exception to this rule: it is instead an example of a *monospaced* font. Text set in Courier closely resembles output from a standard typewriter,[14] because each symbol—letter, number, punctuation mark—claims precisely the same amount of horizontal space.

An interesting font often included as a standard feature with laser printers bears the singular name *Zapf Dingbats*. This reflects its conception by the distinguished contemporary type designer Hermann Zapf, as well as the fact that it consists of a selection of special signs and symbols ("dingbats") uniquely associated with the typesetter's art (cf. Fig. 2–9).

Newcomers to desktop publishing are likely to be content at least initially with the selection of fonts provided with their laser printers, partly for reasons of convenience, but also because high-quality alternative font sets tend to

---

[14] In fact, Courier is based on a design created for IBM's SELECTRIC series of typewriters. The first printer's font purposely created to simulate typewriter output was introduced in 1884 by the Central Type Foundry of Boston at the suggestion of J. C. Blair, a stationer from Huntingdon, Pennsylvania (Kleper, 1990).

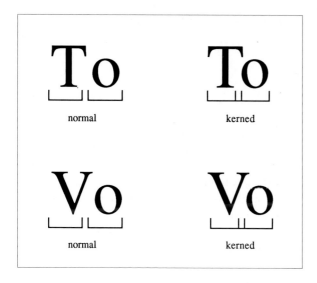

Fig. 2–7. Kerning as a way of improving the spacing of certain letter pairs.

be rather expensive (typically $35–150, depending on the source and nature of the font). Printing is also fastest with "built-in" fonts because these require the minimum amount of data transfer between computer and printer. Indeed, for desktop publishing in the natural sciences, three standard fonts suffice to meet all the basic demands—*Times*, *Helvetica* (Fig. 2–8), and *Symbol* (Fig. 2–10). Times is quite satisfactory for running text, Helvetica takes care of a few special symbols that must be set in sans serif type (cf. Sec. 3.4)—and provides an interesting alternative for boldface section headings—while Symbol (cf. Sec. 3.9) furnishes the Greek letters and many of the special characters required in scientific text.[15] A professional typographer would undoubtedly wince at this generalization, since it so strongly reinforces the often- (and justly-) criticized visual monotony of amateur desk-

---

[15] Times (more accurately: Times New Roman) was developed in 1932 for use by *The Times* newspaper of London. One of its important characteristics, betraying its newspaper origin, is that it permits a large quantity of text to be compressed into a relatively small space. Helvetica originated in 1957 in the Hass typefoundry of Basel, Switzerland, although its roots can easily be traced to typefaces of the late 19th century. Adobe Systems of Mountain View, California, designed the desktop-computer font Symbol.

```
Courier
abcdefghijklmnopqrstuvwxyz!'""$%&/()=?
ABCDEFGHIJKLMNOPQRSTUVWXYZ1234567890
```

Helvetica
abcdefghijklmnopqrstuvwxyz!'""$%&/()=?
ABCDEFGHIJKLMNOPQRSTUVWXYZ1234567890

Avant Garde
abcdefghijklmnopqrstuvwxyz!'"$%&/()=?
ABCDEFGHIJKLMNOPQRSTUVWXYZ1234567890

Kabel Plain
abcdefghijklmnopqrstuvwxyz!'"'$%&/()?
ABCDEFGHIJKLMNOPQRSTUVWXYZ1234567890

Times
abcdefghijklmnopqrstuvwxyz!'""$%&/()=?
ABCDEFGHIJKLMNOPQRSTUVWXYZ1234567890

Bookman
abcdefghijklmnopqrstuvwxyz!'""$%&/()=?
ABCDEFGHIJKLMNOPQRSTUVWXYZ1234567890

Garamond
abcdefghijklmnopqrstuvwxyz!'""$%&/()=?
ABCDEFGHIJKLMNOPQRSTUVWXYZ1234567890

New Century Schoolbook
abcdefghijklmnopqrstuvwxyz!'""$%&/()=?
ABCDEFGHIJKLMNOPQRSTUVWXYZ1234567890

Palatino
abcdefghijklmnopqrstuvwxyz!'""$%&/()=?
ABCDEFGHIJKLMNOPQRSTUVWXYZ1234567890

Fig. 2–8. Examples of several typefaces adapted for use with laser printers.

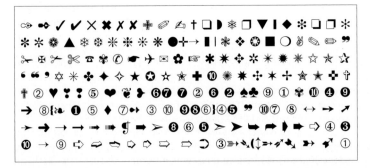

Fig. 2–9. Typographic symbols provided in the font Zapf Dingbats (ITC).

1 2 3 4 5 6 7 8 9 0 − = θ ω ε ρ τ ψ υ ι ο π [ ] ∴
α σ δ φ γ η φ κ λ;Π ζ ξ χ ϖ β ν μ , . / ~! ≅ #
∃ % ⊥ & * ( ) _ + Θ Ω Ε Ρ Τ Ψ Υ Ι Ο Π { } |
Α Σ Δ Φ Γ Η ϑ Κ Λ : © Ζ Ξ Χ ς Β Ν Μ < >
? ℑ ♠ ≤ ′ ° / ƒ ∞ ≈ ... ∠ ↑ ∉ • ↔ ♦ × ← ⊥ ⌡ ≠
® ™ ∩ ♣ ∂ ⊗ ♥ ∅ ⌋ ℜ ⊃ — | ⊕ ℘ ≡ ~ ∝ ≥ √
∧ ∨ ⇔ ⇐ ⇑ ⇒ ⇓ ◊ ϒ ⟨ ® ∇ ± ∈ © ™ Σ ⎛ | ⎝ ⎡
↓ ÷ © Π ∪ | ⎣ ⎧ ⎨ ⎩ | ⟩ ∫ → ⎞ | ·⎠ ⎞ | ⎠ ⎤ ℵ

Fig. 2–10. Character set for the laser-printer font Symbol (Adobe Systems).

top publications, but the proposed three-font combination does satisfy the minimum requirements, its component parts are mutually compatible from an aesthetic point of view,[16] and it has stood the important test of reader acceptance. The beginner is in fact advised to resist excessive typeface experimentation, since serious problems may arise, for example, in the attempt to create a matching set of mathematical equations for a document with body text set in an alternative serif font like Palatino, Bookman, or Garamond.

Despite this admonition, one should not feel absolutely restricted to a minimal font set, or even to the standard fonts supplied with a particular computer or printer. Literally hundreds of high-resolution (laser) fonts are available for purchase. These are supplied in the form of "downloadable" disk (or CD-ROM) files[17] by software distributors, specialized "font houses" (e.g., Precision Type), and such primary sources as Adobe, Agfa/Compugraphic, Casady & Green, Hell-Linotype, and Varityper (cf. also Sec. 7.3).[18]

---

[16] It should be noted, however, that characters of a given size in Times look significantly smaller than the "same size" characters in either Helvetica or Symbol (cf. Sec. 2.4). For this reason it is wise to reduce slightly (e.g., by one point) the size of any characters from Symbol that must be set in close proximity to Times text.

[17] Separate files are usually required to meet the differing requirements of low-resolution screen displays (*screen fonts*, utilized also by dot-matrix printers) and high-resolution (e.g., PostScript) printers or phototypesetters (cf. Sec. 6.3).

[18] Useful sources of additional information on the characteristics and availability of commercial typefaces include Lieberman (1968), Fenton (1991), Grosvenor, Morrison, and Pim (1992), and Lawson (1990), as well as various distributors' catalogues (e.g., the Precision Type *Reference Guide*). An especially comprehensive introduction to the recognition and selection of typefaces is the book by Perfect and Rookledge (1991).

Specific type designs can be propagated not only in a variety of sizes, but also in various *styles*, which combine to form *font families* (e.g., the *Times family*, Fig. 2–2) with varying numbers of members. The principal member of such a family is usually characterized by straightforward upright ("roman") letters, while others may feature alternative line widths (e.g., **Times bold-face**) or axes that have been tilted relative to the baseline (e.g. *Times italic*).[19] Other familiar styles include

SMALL CAPITALS,[20] outline, and shadow.

In typesetting the text of a book or similar work it is always best to restrict oneself to type from a single family except in those few instances where this is impossible, as with certain equations (cf. Chapter 3). With scientific material it is perfectly appropriate to work exclusively with one basic font, such as Times, making only occasional use of style variants (*italic*, **bold,** ***bold italic***) as required for highlighting purposes.

---

[19] Actually, a distinction should be made between a *true* italic font, which is based on an independent design (as first introduced by the Italian master printer Aldus Manutius in 1500; his life work is described by Barolini, 1992, in another beautiful example of desktop publishing), and *slanted* or *oblique* type, in which characters from a roman (upright) font have simply been realigned mathematically in a consistent and aesthetically pleasing way (e.g., *a* vs. *a*).

[20] Small capitals provide an excellent opportunity for even the layman to perceive the difference between a typical piece of computer output and the more sophisticated product of a phototypesetter. A true small capital should have a character height that corresponds exactly to the *x*-height (cf. Fig. 2–13) of the parent typeface, and line-widths for the various letters should have been adjusted to be compatible with ordinary small letters, thereby assuring a consistent level of "grayness" on the printed page. Word-processing and layout programs such as WORD (Sec. 9.4) and PAGE-MAKER (Sec. 10.2) offer resizing algorithms that attempt to simulate the effect of small capitals on the basis of ordinary capitals; e.g.:

TEXT WITH SMALL CAPITALS (in WORD)

in which ordinary capitals have been reduced only slightly relative to the surrounding text, or

TEXT WITH SMALL CAPITALS (in PAGEMAKER).

Though far from perfect, the result in the latter case corresponds more nearly to a typographer's expectations; moreover, the size of the characters here can be adjusted by the user as one component of style definition. The best results are obtained with

A FONT WITH TRUE SMALL CAPITALS

(cf. note 13 on p. 51).

Highlighting is a way of drawing the reader's attention to very special text elements. Points to be emphasized in running text should generally be set in *italic* type, a striking but relatively unobtrusive style that lends itself even to rather lengthy passages. An important characteristic of italic type is that it confers essentially the same degree of "blackness" on a printed page as the corresponding roman font. One should avoid any temptation to resort instead to such alternatives as underlining or capitalization ("all caps"). Underlining is a crude and unattractive device acceptable only in typed manuscripts (where no other choice is available), and long strings of capital letters are rather difficult to read, as noted previously. Blocks of capitals also tend to "disturb the peace" in a paragraph, in part because the characters look even larger than they really are. Indeed, if a situation arises in which a word *must* be fully capitalized (e.g., an acronym like WYSIWYG) it is wise, as here, to set it in slightly (e.g., 10%) smaller type, thereby counteracting the apparent size discrepancy. While full capitalization is strongly discouraged, SMALL CAPITALS (which are much less intrusive) are often recommended as a subtle way of calling attention to the name of an author, for example, or that of a company. Small capitals are also ideal for use as roman numerals, and they can be very effective in headings (cf. Sec. 2.8).

Bold and ***bold italic*** type are "eye-catchers", especially useful for alerting the reader to potential hazards. Boldface type is also well adapted to headings. Its most common application in running text is in conjunction with *paragraph titles*: keywords or phrases placed at the very beginning of certain paragraphs as a way of preparing the reader for a new topic. As noted previously, the contrast provided by boldface type is useful for arresting the reader's attention, but if it is to produce this effect it must be employed sparingly.

The somewhat grotesque outline and shadow styles have no place in serious text, and they will not be discussed further.[21] Another highlighting device, s p a c e d t e x t , was once rather common, especially in German documents, but it has been shown to have a serious negative impact on readability, and has now become virtually obsolete.

Used judiciously, highlighting is unlikely to detract from the overall appearance of a document so long as one stays within the confines of a single type *family*, but problems may arise if families are intermingled (e.g., type

---

[21] One exception to this blanket condemnation is use of the outline style to indicate certain types of mathematical *sets*, as with the $\mathbb{N}$ by which mathematicians represent the set of all natural numbers.

> Used **judiciously**, highlighting is unlikely to detract from
> the overall appearance of a document *so long as* one stays
> <u>within</u> the confines of a **single type** *family*, but problems are
> likely to arise if families are intermingled (e.g., type from a
> *serif* face like Times together with a *sans serif* face such
> as Helvetica). One should in any case avoid mixing too
> many FONTS, STYLES, and SIZES. <u>Excessive</u> variation is
> **unsettling**, and it makes it almost *impossible* for the average
> reader to decipher ***significance*** that might be attached to a
> **particular** form of highlighting.

Fig. 2–11. An example of the chaotic consequences of intermingling too many typefaces and type styles.

from a serif face like Times together with a sans serif face such as Helvetica). One should in any case avoid mixing too many fonts, styles, and sizes. Excessive variation is unsettling, and it makes it almost impossible for the average reader to decipher significance that might be attached to a particular form of highlighting (cf. Fig. 2–11).

An entirely different way to highlight a text element is to make it *freestanding*—to isolate it physically from its surroundings, usually with different (generally wider) left and/or right margins. This device is discussed more fully in Sec. 2.7.

2.4 Units of Measure in Typography

The *size* of a letter or word is in some sense relative: large type seen from a distance obviously "becomes" (is perceived as) small, as in the case of a poster or a billboard. Nevertheless, type-size definitions are important, if for no other reason than as an aid to precise description of a printed work, particularly since characters of different sizes are sometimes used within a single document to suggest the relative importance of various messages.

Many are surprised to learn that specifying a "type size" is actually a rather challenging assignment. It is made especially complicated by the fact that even though two typefaces can be proven to incorporate elements of the "same" size, one face may still *appear* to be significantly larger. Furthermore, there can be a significant difference between type as it is displayed on a computer screen and the "same size" type printed with a laser printer (in part due to the limited resolution of screen fonts). Finally, a set of "10 point" characters from one font may in fact have dimensions markedly different from those of another "10 point" font, as illustrated by the following screen fonts:

Times 10 Point   BostonII 10 Point

Some of the confusion in this area has a historical foundation. The transition from "hot type" (metallic type, as from a Linotype machine) to "cold type" (the basis of photographic typesetting)—which began about 1935—completely reshaped the printing industry, and one consequence was a need to rethink basic terminology and invent new ways of expressing new concepts. In some cases old (technically obsolete) terms were in fact retained, but with revised meanings, and some of these changes relate directly to type size. With "hot type", printing a letter involves transferring ink from the surface of a piece of metal onto a sheet of paper. Each printed character thus

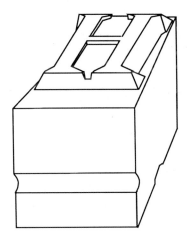

Fig. 2–12. The structure of a typical piece of metallic type.

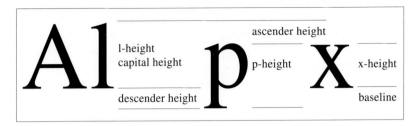

Fig. 2–13. The meanings of the terms "l-height", "capital height", "descender height", "ascender height", "p-height", and "x-height".

originates in the sculpted face of a solid block of lead (cf. Fig. 2–12). A particular *size* was ascribed to such type based on the dimensions of the *block*, not the carved symbol. With phototypesetting, on the other hand, *images* of letters are transferred by optical or other means onto photographic film. In this case the letters can be made any size desired simply by adjusting the projection mechanism.

Describing precisely a set of printed characters in this context requires reference to several different measures, including the *ascender height*, *capital height*, *H-* or *l-height*, *x-* or *mean height*, and *p-* or *descender height* (cf. Fig. 2–13). The *overall* size of a type family is established by measuring the *maximum* vertical extension exhibited by a complete set of letters, which can be determined accurately only by close examination of specific characters like the italic *f* and *Q*. This is often impossible on the basis of a limited sample of printed text, so one ordinarily resorts to *estimating* true type size after measuring some convenient group of several adjacent letters, or one of the capitals, such as H. *Typesetting tables* for the typeface in question permit one to effect an accurate conversion between an observed "H-height", for example, and the corresponding true overall type size. A rough generalized scale for this purpose is included as one element in a device known as a *typometer* or *overlay gauge* (cf. Fig. 2–14), which is a flexible plastic strip typographers use to make a variety of important measurements.

Like typographers, the developers of word-processing and layout software specify type sizes in terms of "points", where the *point* is a unit of measure in the Anglo-American "pica system":

1 point corresponds to 0.351 mm, roughly 1/72 of an inch

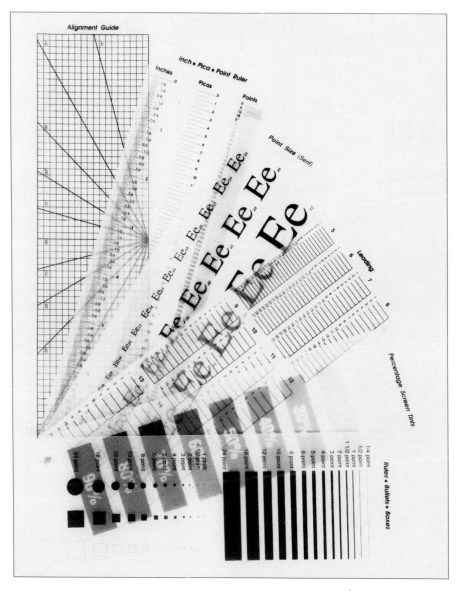

Fig. 2–14. Several "leaves" from a multipurpose set of transparent overlay gauges distributed by Graphics Products Corporation (Wheeling, Ill.) under the name "Graphic Toolkit".

The *point* (or *punkt*) is also a basic unit of measure in another typographic system, the *continental* or *Didot* system, but a Didot "point" has a slightly different value:

$$1 \text{ point (Didot) corresponds to } 0.376 \text{ mm}[22]$$

In other words, "12-point type" implies letters from a set with a maximum extension of 4.21 (or 4.51) mm. The indicated ambiguity reflects the fact that the Anglo-American scale is about 7% smaller than the continental scale. This same unit of measure is also applied to the spacing (*leading*; cf. Sec. 2.5) between adjacent *lines* of type.

The 7% difference between the Anglo-American and Didot scales might seem insignificant, but it can become quite important with large type, or with pages containing many lines. Assume, for example, that one wishes to prepare a page consisting of a single block of text 25 cm (ca. 10″) high, and that this is to be set in 13-point type with a 3-point separation between lines. The Anglo-American system would allow space for 44.5 lines, compared to only 41.5 lines in the Didot system. Such a discrepancy clearly must be taken into account if one happens to be working with foreign software, or dealing with a publishing house in continental Europe. Table 2–1 provides a set of conversion factors for the two standard systems of measure. The following mathematical relationships may also prove helpful (cf. White, 1988):

1 point (A-A) = 0.934 point (Didot) = 0.351 mm = 0.0138 in. (= 1/12 pica)
1 point (Didot) = 1.07 point (A-A) = 0.376 mm = 0.0148 in. (= 1/12 cicero)
1 in. = 67.542 point (Didot) = 72.291 point (A-A) = 25.400 mm
1 mm = 2.659 point (Didot) = 2.846 point (A-A) = 0.0394 in.

Several of the most common type sizes long ago[23] also acquired "nicknames" that were used routinely by printers in the workplace. For example, what we would now describe approximately as 8-point type was once known

---

[22] The usual abbreviation for "point" is "pt" in English, but "p" in other languages. A larger unit, the *pica* (4.216 mm) is equivalent to exactly 12 points. The counterpart to the pica in the continental system is the *cicero* (4.513 mm). The French typefounder François-Ambroise Didot invented the typographic point in 1785, but it was not defined precisely until 1878 by the German typefounder Hermann Berthold in Berlin, who established the relationship:

$$1000 \text{ mm} = 2660 \text{ p (at 0 °C) or } 1000.333 \text{ mm} = 2660 \text{ p (at 20 °C)}$$

[23] Before the invention of the "point" system!

as "Brevier", and 14-point type was called "English" (cf. Table 2–2). Type 12 points in size was referred to as "Pica", a term that soon came to be applied more broadly, so that *any* linear distance of 12 points (i.e., 4.216 mm, ca. 1/6 in.) was referred to as a *pica*. A particular text block might thus be characterized by a printer as "27 × 42", meaning 27 picas by 42 picas (approximately 4.5″ × 7″).

Table 2–1. Table for converting between Anglo-American points, Didot points, and characteristic (overall) heights (in millimeters and inches).

Point (A-A)	Point (Didot)	Height (mm)	(in.)	Point (Didot)	Point (A-A)	Height (mm)	(in.)
1	0.934	0.351	0.0138	1	1.070	0.376	0.0148
2	1.869	0.703	0.0277	2	2.141	0.752	0.0296
3	2.803	1.054	0.0415	3	3.211	1.128	0.0444
4	3.737	1.405	0.0553	4	4.281	1.504	0.0592
5	4.672	1.757	0.0692	5	5.352	1.880	0.0740
6	5.606	2.108	0.0830	6	6.422	2.256	0.0888
7	6.540	2.460	0.0968	7	7.492	2.632	0.1036
8	7.474	2.811	0.1106	8	8.563	3.009	0.1184
9	8.409	3.162	0.1245	9	9.633	3.385	0.1332
10	9.343	3.514	0.1383	10	10.703	3.761	0.1480
11	10.277	3.865	0.1521	11	11.773	4.137	0.1628
12	11.212	4.216	0.1660	12	12.844	4.513	0.1776
24	22.423	8.433	0.3319	24	25.688	9.026	0.3552
48	44.847	16.865	0.6638	48	51.375	18.051	0.7104

Table 2–2. Examples of typographic measures and their traditional (Anglo-American) designations.

Size (pt)	Traditional name	Height (mm)	Size (pt)	Traditional name	Height (mm)
3.5	Brilliant	1.230	12	Pica	4.216
4.5	Diamond	1.581	14	English	4.918
5	Pearl	1.757	16	Columbian	5.621
6	Nonpareil	2.108	20	Paragon	7.028
7	Minion	2.460	24	Double Pica	8.433
8	Brevier	2.811	30	Five-Line Nonpareil	10.542
9	Bourgeois	3.162	36	Double Great Primer	12.649
10	Long Primer	3.514	42	Seven-Line Nonpareil	14.754
11	Small Pica	3.865	48	Canon	16.865

2.5 Line Length and Line Spacing

It would be wrong to assume there is any such thing as an "ideal" line length (or column width), just as there is no universal ideal line spacing. What matters most is overall appearance and readability, and these are qualities that reflect *relationships* between type size, line length, and line spacing. With respect to *length*, lines should be set in such a way that they contain neither "too little" nor "too much" text. When lines are too long the reader is unable to capture a complete line at a single glance, and locating the beginning of the subsequent line becomes awkward. Lines that are too short fragment the text, and generally make inefficient use of the available space. A multi-column layout often produces the best compromise with a large page format (a typical newspaper provides a perfect example). A widely accepted standard recommends limiting the average line of text to roughly 50–70 characters. Another way of expressing the same idea is to say that a typical line should contain no more than about 8–11 words. With most typefaces and type styles, 65 characters per line probably constitutes a good balance.

One of the first rules typesetters learn about *line spacing* is that descenders from one line must never be allowed to touch ascenders in the next (except

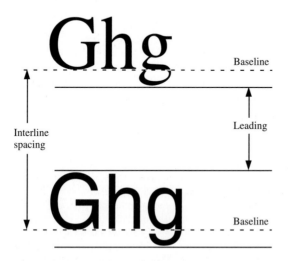

Fig. 2–15. Illustration showing the distinction between "leading" and "interline spacing".

in unusual situations where a special effect is desired—in an advertisement, perhaps). Two important technical terms in this context, *line* (or *interline*) *spacing* and *leading* (pronounced "ledding"), are defined as illustrated in Fig. 2–15.[24] Specifying a particular leading is the usual way of ensuring that there will be sufficient blank space above each line in a block of text to prevent inadvertent contact. A practical reason for introducing leading is that it helps to "fix" a reader's glance during the reading process. Different amounts of leading are preferable in different circumstances, but the leading in a typical book usually corresponds to at least 20% of the associated type size—even more if the lines are unusually long. In other words, 10-point type should ordinarily be set with at least 2 points of leading, which is the same as specifying a line spacing of 12 points. Such an arrangement would be expressed as "10/12", read as "10 on 12".[25]

It is instructive at this point to consider how these parameters relate to *typewritten* text. Typewriters are usually designed to produce characters with one of two sizes, commonly called "pica" and "elite". The smaller of the two, elite, is also known as "12-pitch" type, because lines of the resulting text contain exactly 12 characters to the inch; in other words, every elite character occupies a space slightly more than 2.11 mm wide. With the minimum setting for line spacing ("single-spaced"), adjacent baselines are separated by 4.23 mm, producing copy roughly equivalent to *solid 12-point text*: 12-point characters typeset with no leading. Other options produce lines separated by 6.35 mm (i.e., 1/2-line or 6 points of leading) and 8.47 mm ("double spaced", where each typed line is separated from its predecessor by one blank line, corresponding to an interline spacing of 24 points). These relationships are illustrated in Fig. 2–16, which also provides examples of typeset text in 12-point Times (a proportional typeface, so the character widths vary) with various amounts of leading.

---

[24] The word "leading" originates in the fact that lines of "hot type" were always separated from one another by thin metal strips of an appropriate width. In later years such strips were often made of brass, but originally they were prepared with the softer material lead.

[25] Computer software often uses the term "leading" where "line spacing" would actually be more correct. In PAGEMAKER, for example, *12-point leading* might be selected for a paragraph set in 10-point type. Extra leading is often required in scientific text containing large numbers of subscripts and superscripts. If so, it should be used *consistently*, not simply with the occasional lines that seem to require it.

```
With the minimum setting for line spacing ("single-

spaced"), adjacent baselines are separated by

4.23 mm, roughly equivalent to solid 12-point text:

12-point characters typeset with no leading.  Other

options produce lines separated by 6.35 mm and

8.47 mm. [double-spaced]
```

```
With the minimum setting for line spacing ("single-
spaced"), adjacent baselines are separated by
4.23 mm, roughly equivalent to solid 12-point text:
12-point characters typeset with no leading.  Other
options produce lines separated by 6.35 mm and
8.47 mm. [one-and-one-half-spaced]
```

```
With the minimum setting for line spacing ("single-
spaced"), adjacent baselines are separated by
4.23 mm, roughly equivalent to solid 12-point text:
12-point characters typeset with no leading.  Other
options produce lines separated by 6.35 mm and
8.47 mm. [single-spaced]
```

With the minimum setting for line spacing ("single-spaced"), adjacent baselines are separated by 4.23 mm, roughly equivalent to *solid 12-point text*: 12-point characters typeset with no leading. Other options produce lines separated by 6.35 mm and 8.47 mm. [12/16]

With the minimum setting for line spacing ("single-spaced"), adjacent baselines are separated by 4.23 mm, roughly equivalent to *solid 12-point text*: 12-point characters typeset with no leading. Other options produce lines separated by 6.35 mm and 8.47 mm. [12/14]

With the minimum setting for line spacing ("single-spaced"), adjacent baselines are separated by 4.23 mm, roughly equivalent to *solid 12-point text*: 12-point characters typeset with no leading. Other options produce lines separated by 6.35 mm and 8.47 mm. [12/12]

Fig. 2–16. The effect of varying the leading (or interline spacing) in text set in the typewriter (monospaced) font Courier (top) and the proportional font Times (bottom). All examples are reproduced at 70% of their original size.

It can easily be demonstrated that, for a given typeface and type size, a modest increase in line spacing markedly improves the readability. Moreover, with any particular type size, the longer the printed line the more leading is required to ensure smooth passage of the eye from the end of one line to the beginning of the next. In case of doubt it is almost always best to choose the larger of two alternative leading values.

Footnotes, captions (legends), and tables are usually set in smaller type relative to body text, and with correspondingly less leading. For example, if body text were to be set 11/13, appropriate parameters for the subsidiary elements might be 10/12 or 9/11.[26]

2.6 *Word Spacing, Letter Spacing, and Justification*

Ideally, the space between any two words in printed text should be roughly as wide as the letter "n"; expressed in points this represents a gap equivalent to one-third of the corresponding type size. Narrower spaces are permissible in short lines, while somewhat wider ones are preferable with long lines. *Condensed* (narrow) type demands less space between words, *expanded* (wide) or *bold* type somewhat more.

Most books are typeset in such a way that the text blocks are sharply aligned along both the right and left edges (i.e., *fully justified*), but achieving this arrangement necessitates a certain amount of flexibility with respect to the spaces separating individual words.[27] Indeed, it is rare to find identical word spacings in any two lines within a given paragraph. Particularly with short lines, and in the absence of deliberate hyphenation, fully justified text can actually become quite unsightly. Judicious hyphenation helps, but this is a crutch that must be used sparingly, because text with too many lines ending in hyphens is also unattractive—and difficult to read. One common

---

[26] Compare these suggestions with the specifications adopted for this particular book, as listed in Appendix C.
[27] Letter spacing might also be subjected to some adjustment, but the results are usually much less satisfactory because of the subtle interplay between letter shapes and space. Spacing adjustments are also considered briefly in Sec. 10.2.

rule of thumb asserts that no more than two consecutive lines should be permitted to end in hyphens. Needless to say, any hyphenation introduced must conform strictly to established linguistic rules (if in doubt, check a dictionary!). Many word processors and page-layout programs offer the option of automatic hyphenation, a subject to which we return in Section 9.2.

Even with hyphenation, full justification may result in unsatisfactory word spacings, and it can impose an air of artificial rigidity upon a block of text, particularly text set in narrow columns (cf. Fig. 2–17). An alternative worth serious consideration is the *left-justified* or "ragged-right" format (also shown in Fig. 2–17), in which alignment is maintained only along the left, and individual lines vary somewhat in length. Word spacing in this case can be held constant, and very little hyphenation is required. While left-justified text does present a somewhat "restless" appearance along its right edge, the overall impression may actually be one of greater uniformity and compactness. Left justification is almost obligatory in the case of extremely short lines (as with marginal notes, for example).

If left justification is to be the rule in the main body of a document, then other elements such as footnotes, headings, and figure captions should also be justified only on the left. On the other hand, *centering* may be a viable option for subsidiary elements if the body text is fully justified. Centered text is otherwise rarely encountered in publications in the natural sciences, though it plays an important role in the layout of display copy and verse. *Right justification* is even more unusual, but it can be very effective with figure captions that are placed to the left of the corresponding figures, since alignment on the right in this case helps "bind" the caption visually to the figure (cf. White, 1988).

One other text arrangement occasionally found in scientific documents, especially in the vicinity of small illustrations or tables, is *wrapped* text, where words are allowed to "flow" around intervening elements. Wrapped text also embraces the "drop capital"[28] sometimes introduced at the beginning of a paragraph (including this one, although drop capitals are usually associated only with paragraphs that introduce new sections). Computer-generated text can even be aligned in such

---

[28] "Drop capitals" are capital letters that have been artificially enlarged in such a way that they extend downward across two or more lines of text (seldom more than five). They are the legacy of a proud typographic tradition, and direct descendants of the magnificent hand-lettered initials in manuscripts from the Middle Ages.

Most books are typeset in such a way that the text blocks are sharply aligned along both the right and left edges (i.e., *fully justified*), but achieving this arrangement always requires a certain amount of flexibility with respect to the spaces separating individual words.

Most books are typeset in such a way that the text blocks are sharply aligned along both the right and left edges (i.e., *fully justified*), but achieving this arrangement always requires a certain amount of flexibility with respect to the spaces separating individual words.

Most books are typeset in such a way that the text blocks are sharply aligned along both the right and left edges (i.e., *fully justified*), but achieving this arrangement always requires a certain amount of flexibility with respect to the spaces separating individual words.

Most books are typeset in such a way that the text blocks are sharply aligned along both the right and left edges (i.e., *fully justified*), but achieving this arrangement always requires a certain amount of flexibility with respect to the spaces separating individual words.

Fig. 2–17. Examples of ragged right, centered, and fully-justified text, the latter both with and without hyphenation. Shading has been used to highlight a passage with excessive hyphenation.

a way that it follows a curved path (cf. Fig. 2–18), but this is a technique rarely seen outside the realm of advertisements, logos, and the like. Typeset curved text would have been almost impossible to achieve only a few short years ago even for professionals, but this limitation disappeared with the advent of powerful graphics software.[29]

---

[29] Among the programs that support curved text are ADOBE ILLUSTRATOR, ALDUS FREEHAND, and CANVAS (cf. Chap. 11).

Fig. 2–18. Text arranged to follow curved and irregular paths, prepared with the aid of graphics software (ADOBE ILLUSTRATOR).

2.7 Indentation, Isolation, Widows, and Orphans

It is customary to *indent*—i.e., inset slightly—the first line of every full paragraph (except the first paragraph in a section) as a way of signaling to the reader that a new thought is about to be introduced. The extent of the indentation is subject to design considerations, but it should never be less than the width of the capital letter "M" (i.e., ca. 4.5 mm in the case of 12-point type). Less common is the practice of marking paragraph breaks with extra leading, in which case indentation becomes superfluous because the vertical offset alone serves as an adequate signal. Whatever approach is selected it should be followed systematically throughout a particular work.

Section 2.3 introduced *highlighting* as the usual way of using typography to emphasize certain words. An important technique for calling attention to a somewhat larger text element is to make it *freestanding*: divorced from the surrounding text and thereby transformed into a separate block with its own distinctive white border. The resulting blank regions serve not only to call attention to the isolated element, but also to confer an extra level of structure on the page, making it visually more interesting. The optimum amount of extra space above and below an isolated text element (e.g., an equation) is normally either half a text line or one full line. (Isolation as it applies to figures and tables is discussed in Sec. 2.10.) Assuming paragraphs have been indented, the left offset for a freestanding element should be at least twice the paragraph indentation. Extra space above (below) a free-standing element is unnecessary—and inappropriate—at the top (bottom) of a page. For specific examples and further suggestions see Section 3.7.

Typesetters consider it a sin to allow any page to begin with the last line of a paragraph or section. The effect is particularly unsatisfactory in the case of a left-hand page, since the resulting "orphan" loses all contact with the main body of the paragraph to which it belongs. Similar considerations apply to a "widow": a first line of a paragraph isolated at the bottom of a page. In other words, one should try to ensure that each page begins and ends with at least two lines of related text.[30]

The same principle applies to headings: a new heading should always be accompanied on the same page by at least three lines of subsequent text. If this is not feasible, then the heading should be moved to the top of the sub-sequent page even though this will leave an unusually large space at the end of the preceding page. In no case should one attempt to "fill" such a truncated page by the artificial insertion of extra blank lines, or by increasing the leading or the space between paragraphs. By the same token it would be wholly inappropriate to solve the problem by arbitrarily setting the offending paragraph in smaller type (or with reduced leading) as a way of compressing it into a space that would otherwise be too small. It is far better simply to accept the fact that one page in the document will have an unusual amount of space at the bottom.

It is often possible to deal with problematic "widows" or "orphans" by examining nearby text for places where unnecessary words or phrases can be eliminated, or an extra (useful!) phrase added. Alternatively one might look for a place (even several pages earlier) where space above and below a freestanding element could be adjusted slightly. A manuscript with many equations presents a golden opportunity for such minor adjustments, and with most professional typesetting equipment they would be made automatically. As an example, consider a document with 16-point line spacing in which an extra 8 points (i.e., 1/2 line) of leading has been introduced before and after each freestanding equation. If one were to increase by one point the space

---

[30] This subject is discussed in greater detail in, for example, *The Chicago Manual of Style* (1982).

above and below each of eight such equations the change would almost certainly go unnoticed, but it would suffice to displace the end of the text downward by the equivalent of a full line, while the reverse process would allow one to recover an additional line of space.

2.8 Chapter and Section Headings

Long documents should always be divided into *chapters*, and long chapters are often divided into *sections* and *subsections*. Organization in this sense should never be carried to extremes, however. Four levels of structure should be sufficient for presenting even a very complex argument; indeed, one should try to avoid dividing a document into sections at more than four levels. Authors are increasingly adopting the *decimal* notation as a way of designating section breaks at the first three levels; e.g.:

6	References in a Scientific Manuscript
6.1	Documentation
6.2	Literature Citations
6.2.1	Numerical Systems
6.2.2	The Name–Date System
6.3	Journal References
6.4	Book References
etc.	

Numbering can be dispensed with at the fourth level *(4th-order subdivisions)*, relying instead on freestanding (or *displayed*) *unnumbered headings* (which may or may not be included in the table of contents), or *paragraph titles* (highlighted words set at the beginnings of paragraphs). The reason for discouraging higher-order numbering is that it tends to confer too much "academic formalism" on a work, often to the point of intimidating the reader and diminishing overall readability.

A new sublevel of text should be created only if it will contain at least two sections. In other words, one must prevent the occurrence of a situation like the following:

...

3.2	Properties of Fluorine-Containing Compounds
3.2.1	Perfluoroalkanes
3.3	Properties of Chlorine-Containing Compounds

...

A potential problem of this type can be averted in one of two ways: either by breaking the material in question into two subsections, e.g.,

...

3.2	Properties of Fluorine-Containing Compounds
3.2.1	General Considerations
3.2.2	Perfluoroalkanes
3.3	Properties of Chlorine-Containing Compounds

...

or, if the subject matter will support the change, by promoting the "lonesome" heading to the next higher level:

...

3.2	Properties of Fluorine-Containing Compounds
3.3	Perfluoroalkanes
3.4	Properties of Chlorine-Containing Compounds

...

In a *typewritten* manuscript, first-order headings are usually distinguished by setting them entirely in capitals. Capitalization in second-order headings is then limited to the first letters of important words, but the entire heading is underlined. Underlining is dispensed with in third-order headings, leaving these to stand out only by virtue of the fact that they are physically isolated from the body of the text. This is clearly a "low-tech" solution, however; while it does convey a set of distinctions, it is not attractive, and it is unworthy of the author with access to a modern word processor. The book publisher's approach is preferable: chapter headings set in larger and/or bolder type than that used for section headings, which are in turn larger than subsection headings. This is the system we have adopted, as illustrated by the upper part of Fig. 2–19.

An alternative (though less satisfactory) way of distinguishing heading levels is to maintain a constant type *size* but vary the type *style*. Thus, headings at the lowest level (which are unnumbered) might be set in normal type, with those next higher in italics, primary section headings in small capitals, and chapter titles in full capitals (cf. the lower part of Fig. 2–19).

Special attention should be devoted to the *spacing* of headings. A convenient rule applicable even to simple reports states that the amount of free space before and after a heading should be greater, the higher the level of the heading, while the *pair* of spaces above and below a heading should reflect at least approximately the ratio 3 : 2.

1 Chapter Title (in 24 point)

1.1 Section Heading (in 18 point)

1.1.1 Subsection Heading (in 15 point)

Unnumbered Heading (in 13 point)

Text text text text text text text text text text text text text text text text text text text text text text text text text text text text text text text text text text text text text text text text text text text

---

1 CHAPTER TITLE (IN CAPITALS)

1.1 SECTION HEADING (IN SMALL CAPITALS)

1.1.1 Subsection Heading (in italics)

Unnumbered Heading (in normal type)

Text text text text text text text text text text text text text text text text text text text text text text text text text text text text text text text text text text text text text text text text text text text

Fig. 2–19. Two alternative approaches to distinguishing between headings at four levels.

2.9 Running Heads

A *running head* is a line of information set above the body text on most pages of a book or similar document.[31] Running heads are usually separated from the text itself by the equivalent of one-half to two lines of text—and perhaps distinguished further by a thin horizontal line. The purpose of a running head is to assist the reader in search of a particular portion of a document.

Running heads can be used to convey various types of information. Some consist of nothing more than the page number *(pagination)*, which is usually set close to the outside edge of the page, perhaps flush with the margin, or indented by some fixed amount. *Even* numbers are always associated with left-hand pages, *odd* numbers are on the right. Pages within scholarly or scientific text usually feature what is called a "living head". This provides in addition to pagination some explicit reference to the text itself: information such as the title of the current chapter or section, the name of the relevant author (in a multi-author or collective work), or, in the case of an alphabetically arranged encyclopedic volume, one or more keywords from the page in question. In some books the running heads on right and left pages are equivalent, but more often, as in this book, right-hand pages indicate the number and title (perhaps abbreviated) of the current *section*, while those on the left display the *chapter* title. This seemingly straightforward system poses one potential hazard for the desktop publisher, however: running heads on the right change quite frequently, so they cannot be incorporated into the master pages, making them a likely site for errors.

Personal preference is a legitimate consideration in the choice of style and content for running heads, but the decision should also be influenced by the nature of the work. Running heads are an important topic for discussion with one's editor in the course of planning for a camera-ready publishing project. Running heads of various types are illustrated in Fig. 2–20 together with a few lines of the subsequent text.

---

[31] Running heads are omitted from the front matter and the first page of each chapter—which should normally be a right-hand page.

940   *B. A. Fowler and P. L. Goering*

extraction, ion exchange or pyrometallurgic processes, followed by raffination
(WIESE, 1977). World production of indium was about 50 tons per year in 1982 and
the main producing countries are the United States, the Union of Soviet Socialist

II.13 Indium   941

LEACH et al. (1961) reported similar findings in rats exposed to In_2O_3 via inhalation. Other investigators (MORROW et al., 1958) estimated the absorbed dose of ^{114}In sesquioxide particles at between 3 and 6% of the total dose following in-

326   Electrophoresis   Vol. B 5

standard techniques. The molecular form in
which the proteins are separated depends strongly on the presence of additives, such as urea and/
or detergents. Moreover, supramolecular aggre-

Since the isoelectric state involves a minimum of
solvation and thus of solubility for the protein
macroion, there could be a tendency for some
proteins (e.g., globulins) to precipitate during

Vol. B 5   Electrophoresis   327

(e.g., the blue dye Remazol) or fluorescent, such
as dansyl chloride, fluorescamine, O-phthaldialdehyde, and MDPF (2-methoxy-2,4-diphenyl-

tion, for measuring molecular size. This has been
accomplished by overcoming charge effects in
two main ways. In one, a relatively large amount

184   *7 Microscopy and Related Techniques*

If the tunneling current is from the surface to the tip, the STM images the density
of occupied states. If the potential is reversed, the current flows in the other direction,

7.3 Scanning Probe Microscopy: AFM and STM   185

test sample for demonstrating that an STM achieves atomic resolution, illustrates
that one must be careful in translating STM pictures directly into ball models of

Fig. 2–20. Examples of various types of running heads, illustrating the treatment
of both left and right pages.

2.10   Figures, Tables, and Footnotes

Figures and tables play a rather complicated role in a document. They are usually perceived by the author as an integral part of the overall intellectual message, but readers often try to acquire an initial sense of the message as a whole from these elements alone. It is for this reason that much careful thought should be devoted to the preparation of table and figure *captions* (cf. Ebel, Bliefert, and Russey, 1987).

Every figure and every table included within a document must be "anchored" by at least one explicit reference in the text. Typical references of this type include:

> ... Fig. 2–10 illustrates ...
> ... as is indicated in Fig. 3–4.
> ... (see the third column of Table 7–1) ...

The corresponding element itself is then incorporated into the layout at a point as close as possible to its first mention (perhaps at the end of the paragraph to avoid interrupting the reader's train of thought). Tables or figures are usually isolated from text above and below by space equivalent to two full text lines. As with equations, extra space is unnecessary above (below) a table or figure at the top (bottom) of a page.

Unless there is an unusual reason for doing so, one should *not* position all figures and tables at the same vertical level. Instead they should be distributed in such a way that individual pages acquire a balanced look. On the other hand it *is* customary to maintain uniformity in the *horizontal* dimension, with all elements centered, perhaps, or aligned against the left margin, as in this book. Figure captions are normally placed *below* the corresponding figures, although placement beside the figures is also a possibility, especially if it results in the saving of space.

Each *figure* should be provided with a carefully conceived explanatory *caption* (sometimes referred to as a *legend*), in part (as noted above) because many readers examine illustrative material before reading the supporting text. Figure captions, as well as footnotes, table entries, and table headings, should be set in type 1–2 points smaller than that used for body text (cf. also Sec. 2.3).

Tables warrant much closer attention than they are often accorded. Many factors contribute to the ease with which a table can be interpreted. Clarity in a complicated table can often be improved by

- adding a few dividing lines (cf. also Sec. 2.12)
- adjusting the way the total space or surface area is apportioned
- distinguishing one particular column from the others by bolder type
- increasing the free space between lines or columns[32]

Once a satisfactory table *style* has been adopted, every effort should be made to follow it consistently throughout the work (although the unique content of certain tables may mandate exceptions). Like figures, tables also must be provided with well-conceived explanatory captions.

When creating tables with a word-processor or page-makeup program one should always rely on the *tab* function rather than blank spaces as a way of achieving alignment. Alignment on the basis of discrete spaces is awkward at best, especially with a proportional typeface, and the problem is compounded by the fact that spacing on a computer screen inevitably differs somewhat from that observed in printed copy (a consequence of differing resolutions for screen and printer; cf. Sec. 6.2). Moreover, if a need should arise later to alter the original typeface or type size, or to add information, a set of columns aligned on the basis of blank spaces would quickly degenerate into a disaster. Tab settings, by contrast, are extremely precise, and they are easily subjected to revision at any time. Modern software offers the added convenience of tabs for right, left, and center alignment, as well as special tabs that automatically align a column of figures along the decimal point (cf. Sec. 9.2).

Notes (or *footnotes*) that may be associated with a table are different from notes elsewhere in the text, and they must be labeled differently. Thus, if text notes are designated by symbols, such as

$$*, \dagger, \ddagger, \S, **, \dagger\dagger, \text{etc.}$$

then table notes should instead be referred to by superscript numbers ([1], [2], [3], etc.) or letters ([a], [b], [c], etc.).[33]

---

[32] Note, however, that optimal table structure is a function primarily of a table's *content*, not aesthetic considerations. Concrete suggestions regarding table design are provided in Ebel, Bliefert, and Russey (1987).

[33] A clear distinction must of course also be maintained between table notes and literature citations.

In some books, including this one, *text notes* are set as *footnotes* at the bottoms of the corresponding pages. Their purpose may be (as here) to provide supplementary information or interesting observations related to the text itself, but in some publications the footnotes supply essential literature references. In the latter case it is strongly advised that the notes be designated *numerically* rather than by symbols, in part because this makes it easier to deal with subsequent references to a previously cited source.

If references are to be collected at the end of a work, or at the ends of chapters, there are strong arguments in favor of utilizing the *name–date system* (as in this book), which eliminates any need for reference *numbers*. Should a numerical system be adopted, however, sequential citation numbers should be set in normal-sized type directly on the text line and enclosed in square brackets, since this convention results in numbers that are both clearly legible and easy to locate (for more on the subject of references see Ebel, Bliefert, and Russey, 1987).

2.11 Punctuation

It may seem odd that we felt it necessary to devote a separate section to the subject of punctuation. We do so purely on typographic grounds: certain punctuation marks should always be set flush with adjacent text, while others should be preceded (or followed) by a single space, and a few are best accompanied by special narrow (*thin*) spaces. Unfortunately, word processors and layout programs rarely provide straightforward facilities for the latter situation.

● *No* space should be left between the following punctuation mark/letter pairs:

   x.   x,   "x   x"   x)   x]

● A *thin* space (ca. 1–1.5 pt) is recommended in four instances:

   x: x;   x?   x!

- *One* full space should be left after a comma, or after a period at the end of a sentence (*not* two spaces in the latter case, as is the custom with *typewritten* copy):

  x. Xxxx   x, xxx

- *One* space is also appropriate before and after the ellipsis points used to signify the omission of text within a quote:

  xxx ... xxx

  Ellipsis points at the end of a sentence are followed directly by a period.

- If all the text enclosed within a set of parentheses is italicized, then the parentheses themselves should be set in italics as well, but punctuation that follows should be roman. If only a portion of the parenthetical text is italicized, the accompanying parentheses are set in ordinary roman type:

  (xxx xxx xxx)! xxx ...
  (xxx *xxx* xxx)
  (*xxx* xxx xxx)
  (xxx xxx *xxx*)
  (*xx* xxxxx *xx*)

The subject of *dashes* warrants somewhat more extensive comment. A typewriter provides only one "dash-like" symbol, the hyphen (-), but professional typesetters routinely make use of at least two others as well: the "en-dash" (–), which is slightly longer (i.e., approximately as long as the letter "n" is wide), and the even longer "em-dash" (—). The longest of the three, the *em-dash*, is used mainly as a delineator for isolating parenthetical comments,[34] and it is most often set flush with the adjacent text on both sides (though some typesetters prefer to leave a small amount of space); i.e.:

  When we met last week—you may recall that it was on Thursday—we decided ...

In typewritten manuscripts an em dash is usually simulated by two consecutive hyphens.

---

[34] Some typographers prefer to use a spaced en dash for this purpose: "... the em dash is too long for use with the best typefaces. Like the oversized space between sentences, it belongs to the padded and corseted aesthetic of Victorian typography." (Bringhurst, 1992)

There are several situations that call for use of the somewhat shorter *en dash*. The most important of these are

- between numbers representing the limits of a *range* (2–10 mm, 1986–89, etc.)
- as a link between *parallel nouns*, including sets of proper names (e.g., acid–base titration, the Diels–Alder reaction)
- as a way of expressing the mathematical *minus sign* (the temperature was $-3\,°C; -4x + 3y - 7z$)

For purposes of clarity it is sometimes helpful to separate a minus sign from the symbol to which it applies by at least a small amount of blank space (cf. Sec. 3.5).

Finally, the *hyphen* is used to signal word division at the end of a line, and as a connecting link in certain compound words (e.g., "mass-produce", "kilowatt-hour").

Many personal computers now offer facilities for the straightforward introduction of all three types of dashes. With the Apple MACINTOSH system, for example, the key combination "option-hyphen" generally produces an en-dash, whereas "shift-option-hyphen" leads to an em-dash. Further information on the subject of punctuation in general—some aspects of which are highly controversial—is available from a wide variety of sources, one of the most authoritative and convenient being *The Chicago Manual of Style* (1982).

2.12 *Lines*

Straight lines (or *rules*) are sometimes added to a document as a way of emphasizing the structure of the text and facilitating its interpretation. One primitive example is the line used for *underscoring* a text element that seems to require highlighting, although lines of this kind are rare in typeset copy— and usually limited to a typewriter-like font such as Courier. As noted previously, typesetting offers far more attractive highlighting alternatives, including italic or boldface type (cf. Sec. 2.3), or isolation of the text element in question (Sec. 2.7). Nevertheless there are other situations in which typeset lines are quite appropriate, or even obligatory. For example, *tables* ordinarily require at least three such lines, all horizontal and of equal length: a pair of lines bracketing the column header (to distinguish it from the table's content), and one at the very end, immediately preceding any table footnotes. Certain tables also incorporate vertical lines to distinguish the columns more clearly, or as a way of isolating specific *blocks* of data.

Footnotes set at the bottom of a text page (to provide supplementary information or literature references) are usually separated from the body of the text by a thin horizontal line. Such a line should be at least 15 mm long, though it might also be extended across the entire width of the text block above.[35]

Printed lines can vary widely in their thickness (or *weight*). In the days of "hot type", line weights were always expressed in points, although descriptive names were occasionally assigned to specific line weights as well (e.g., a *fine* line traditionally had a weight of 1/5 point). Even today, the finest line available may still be referred to as a *hairline*,[36] but line weights are more often expressed in millimeters. Many word processors, and most layout and graphics programs, include special facilities for introducing lines of various weights and styles (cf. Table 2–3).

Table 2–3. Examples of typographic lines.

Weight	Designation	Example
0.088 mm	1/4-point rule	————————
0.176 mm	1/2-point rule	————————
0.351 mm	1-point rule	————————
0.703 mm	2-point rule	————————
	Double rule	————————
	Scotch rule	————————
	Coupon rule	— — — — — —
	Leader	· · · · · · · · · · · · ·

[35] Full-length dividing lines are occasionally reserved for signaling the fact that footnote text has been *continued* from the preceding page.
[36] The vagueness of this definition ("the finest line available") is one reason for avoiding the "hairline" option in a word-processing or layout program.

2.13 The Finished Product; Choice of Paper

In Sec. 2.1 we observed that typography in the broad sense affects almost every aspect of a printed work. We then proceeded to limit our attention to a much narrower interpretation of the subject. Before concluding, however, it seems appropriate to consider briefly the finished product of a typographic endeavor, and especially the material form it should take.

Once the electronic precursor to a physical document has been completed, several alternatives are available for transforming it into a publication.

● In the case of a small print run it is often most convenient simply to prepare an appropriate number of *photocopies* from a laser-printer original. A simple binding might then be added in one's own office or at a copy shop, although a more elaborate binding could be created by a professional bookbinder.

● Another viable approach to a relatively small number of copies is use of a photocopying device to produce a corresponding set of *offset masters*, which are subject to duplication with relatively simple and inexpensive printing facilities.

● True *printing plates* might also be prepared from laser-printer originals by a special photographic process. The resulting plates could be of the same size as the originals, but *reduction* and *enlargement* are also possible.[37] The plates themselves would then be processed further by a professional printer.

● Finally, one could prepare a set of high-resolution *films*, again at any desired size—but this time directly from the computer file containing the original electronic document, not from laser prints. Such films would of course also require further processing by a professional printer. Analogous copy generated with a laser printer would in this case represent only an approximation of the final product, although a set of preliminary laser

---

[37] Photoreduction of camera-ready copy has the advantage that it leads to copy with a higher degree of apparent resolution (cf. Sec. 6.2). When a printer speaks of a "reduction to 50%" what is often meant is reduction of the *total print area* by 50%, corresponding to a *linear reduction* (i.e., of the height or width) by the ratio $1:\sqrt{2} = 1:1.414$ (to 70.7% of the original value).

prints is indispensable for verifying the general suitability of a proposed layout.[38]

Once a particular production method has been selected, the next step is to choose the right type of paper—whether for the finished product or for camera-ready copy. Here it is always best to seek advice from a professional—perhaps after first acquiring a bit of background from a source like the excellent book by Beach and Russon (1989). The nature of a paper's *surface* plays a significant role in print quality, since this controls the way that ink (or toner) is deposited and distributed. The surfaces of many printing papers are subjected by the manufacturer to special *calendering* or *coating* processes to alter their characteristics.[39] The thickness and density of the paper are also important parameters. Different grades of paper in the United States are distinguished on the basis of the total mass (expressed in pounds) of *500 sheets* of paper cut to one of the so-called *basis sizes*: $25'' \times 38''$ for book paper, $17'' \times 22''$ for business paper. Everywhere else in the world paper is instead classified according to its *surface weight* (in g/m^2). Laser-printer output intended as camera-ready copy, especially if it includes half-tone illustrations, should be prepared on at least 24 lb (ca. 90 g/m^2) stock.[40]

The optimum paper *format* for a publication is to some extent a function of the intended purpose. Most scholarly books are designed around a *verti-*

---

[38] It is important to emphasize that even though laser-printer copy will *resemble* the final product—probably very closely—the two will *not* be identical, and the possibility of unpleasant surprises cannot be ruled out. This subject is discussed further in Sec. 7.5.

[39] *Calendering* is a finishing step in the papermaking process in which the paper stock is passed between one or several pairs of polished steel rollers moving at different rates or even in opposing directions, thereby causing a polishing effect. A *coated* paper is one to which a casein-containing chalk layer has been applied, leading to a surface that can be either more or less flat, and either "matte" or "glossy".

[40] *Metric* papers are the standard elsewhere in the world. They offer the great advantage of a rational and consistent set of size and shape relationships. For example, an international letter sheet, designated "A4", is 210 mm $\times$ 297 mm—slightly longer but also narrower than US letter paper (which is about 216 mm $\times$ 279 mm). If such a sheet is divided in half it gives two smaller (A5) sheets whose relative dimensions are identical to those of the original ($1:\sqrt{2}$, roughly 5:7), and this process can be repeated indefinitely with comparable results. The system is based on a full-sized sheet (A0) with the dimensions 841 mm $\times$ 1189 mm and a total surface area of 1 m^2.

cal (or *portrait*) format. Common book dimensions in the United States include 5″ × 7″ and 6″ × 9″. Given a choice, one should always try to select a page size that can be created from standard commercial paper stock with a minimum amount of waste.

3 Typesetting for Mathematics and the Natural Sciences

3.1 Introduction: The Special Challenges Posed by Scientific Text

Successful typesetting of scientific text requires mastery of a surprising number of rules and principles, most of them designed to assure clarity and the unambiguous interpretation of symbolic expressions, which might be mathematical, physical, or chemical. Scientists benefit from these rules every day as they pore over the published literature, but few are aware even of their existence. In a sense this is as it should be, because good typography rarely calls attention to itself. Nevertheless, the desktop publisher trying to confer form on a piece of scientific text must be more alert, because modern computer-generated copy is expected to reflect considerably more sophistication than a typewritten manuscript—and to convey more information. For example, a typist normally has access to only a single typeface, but the wide range of typefaces and type styles one can generate with a computer opens the way to important typographical distinctions, as between the various functional components in a mathematical expression. A typewriter equipped with a removable type ball or some system for introducing special characters one at a time makes it possible in principle to take advantage of typographic subtleties like italics and boldface, or to introduce occasional alternative symbols, but the effort entailed is virtually prohibitive. On the other hand, the scientist with access to a good personal computer, the right software, and a high-resolution printer has all the tools required to prepare professional-quality camera-ready copy for the most complex technical document.

Exploiting these tools to their full potential represents a significant challenge, however, and it requires a conscientious commitment. The first step is becoming conversant with internationally sanctioned typesetting conventions as the only way of avoiding embarrassing and even costly mistakes.

Specific rules have been developed for distinguishing various types of *symbols*, for example, including those for physical quantities, units, variables, vectors, and operators, requiring application at the appropriate times of italic, boldface, and even sans serif type (cf. Sec. 2.3), and other rules set standards for *spacing* and character *size*. Many of the rules have been codified in the form of national and international *standards*, including standards developed and promoted by the American National Standards Institute *(ANSI)* and the International Organization for Standardization (*ISO*; see the bibliography for further information). A convenient summary of many of the preferred symbols of "scientific shorthand" is the International Union of Pure and Applied Chemistry *(IUPAC)* publication *Quantities, Units, and Symbols in Physical Chemistry* (1988), which also contains a brief discussion of typographic conventions.

The essential elements of scientific typesetting will become apparent in the course of a careful consideration of the correct presentation of *formulas* of various kinds. Formulas present a number of typographic challenges, not least of which is the need for complete flexibility with respect to the vertical placement of specific characters. A complex mathematical expression cannot be forced to conform to standard line-spacing parameters. Careful placement of individual formula elements is essential for ensuring that readers will be able to interact efficiently and simultaneously with all the symbols present, and also for preventing ambiguity regarding the precise role of any particular symbol. *Sizes* of the various characters in a formula must also be appropriate: while some may quite properly be the same size as characters in the surrounding text, the context may require others to be set significantly smaller (or larger!).

Generally speaking, the symbols constituting a particular formula should be distributed in a balanced way with respect to a single reference line, the *formula axis*, which often contains operation symbols like "+", "–", "×", and "=". If possible, the formula axis should be so positioned that it coincides with what would otherwise be one of the text lines on a standard page. It is also common practice to *isolate* complex mathematical, physical, and chemical expressions from the main body of the text, causing them to appear as *freestanding* or *displayed* entities (cf. Sections 2.7 and 3.7).

We conclude this brief introduction with an example of the irresponsible typesetting too often encountered in the work of amateur scientific desktop publishers. The following mathematical expression was included in an in-house publication from a respected scientific company:

$$E_T = \left(\frac{B_{ij}}{d_{ij}^{12}} - \frac{A_{ij}}{d_{ij}^{6}} \right) + \frac{\sum U_n \cos(n\tau + \delta_n)}{2}$$

The "typesetter" responsible for this particular piece of work was clearly unprepared for the assignment. Notice, for example, that no attempt has been made to organize the symbols around a common axis. While the correct symbol has been used to represent a summation, it is far too small, and it is even in the wrong place, leading to serious ambiguity (should the "2" be regarded as falling within the summation?). Finally, all elements in the expression—not only operations symbols but also variables and quantity symbols—have been set uniformly in roman type, which is clearly inappropriate. A much more satisfactory representation of what the author (presumably) had in mind would be:

$$E_T = \left(\frac{B_{ij}}{d_{ij}^{12}} - \frac{A_{ij}}{d_{ij}^{6}} \right) + \sum \frac{U_n}{2} \cos\left(n\tau + \delta n\right)$$

In subsequent sections of this chapter we will return for a closer look at several specific points only touched upon here. First, however, a slight digression is in order.

3.2 Learning From the Past: The Traditional Approach to Scientific Manuscripts

Before we embark on a rigorous analysis of scientific typesetting in a desktop publishing context it will be worthwhile to consider briefly the traditional approach to publishing, especially the editorial process and the crucial "markup" step. "Marking" (or "marking up") a manuscript is quite different from "revising" a document. *Revision* is an activity an author engages in during the early stages of manuscript preparation as a way of strengthening the *content* of a document. *Marking* is one of the last steps in preparing a completed manuscript for *typesetting*. The object is to provide a clear set of instructions to guide the typesetter, particularly with respect to text elements that require special treatment (e.g., a special typeface, unusual margin settings, space to be deliberately set aside for a particular line drawing, etc.).

The final draft of any manuscript should be structured in such a way as to facilitate—not impede—the subsequent editing process, an admonition just as applicable to a manuscript that will be "electronically typeset" by the author as to any other. Pages should all be of standard letter size, for example, and text elements should be uniformly double-spaced and surrounded by wide margins—especially on the right. Some publishing houses provide their authors with special manuscript paper that includes pre-printed guides for a 1-inch margin on the left (just enough to permit binding), but a right margin of as much as 3 inches, leaving sufficient space for roughly 50 typewritten characters per line, with a total of 28–30 lines per page.[1] Each manuscript page must of course be clearly and unambiguously numbered, preferably according to a system that also identifies the current chapter (i.e., "3–6" to indicate the sixth page of the third chapter).

Last-minute wording changes can be inserted (clearly!) by hand, preferably between the lines and *above* the passages to be amended. Wording changes suggested later by the editor will be indicated in a similar way.[2] If necessary, more extensive changes can be entered in the margins, but in this case appropriate links to the pre-existing text must be made exceptionally clear (cf. Fig. 3–1).[3] Careful work is essential, because it may be that a secretary without scientific training will at some point be asked to type a "clean" copy of the manuscript, or to enter the author's corrections into an existing computer file.[4]

The first responsibility of an editor receiving such a manuscript is to examine it and annotate it wherever necessary with requests for clarification,

---

[1]   Standard manuscript paper is made available not as an act of charity, and not simply to facilitate editing, but also because its use makes it easier for the publisher to estimate in advance the likely extent of a particular work once it has been recast in its final printed form.

[2]   The latter are of course regarded as tentative and subject to the author's approval (or at least to negotiation).

[3]   Despite the emphasis here on clarity, changes at the manuscript stage can be handled quite informally; in particular, they are not expected to conform to formal *proofreading* conventions.

[4]   Indeed, the ease with which one can easily prepare clean copies of a manuscript—often several times during the drafting and editing process—is one of the great benefits of computer-based writing. It also increases the probability that authors will continue polishing their work until it is truly "finished".

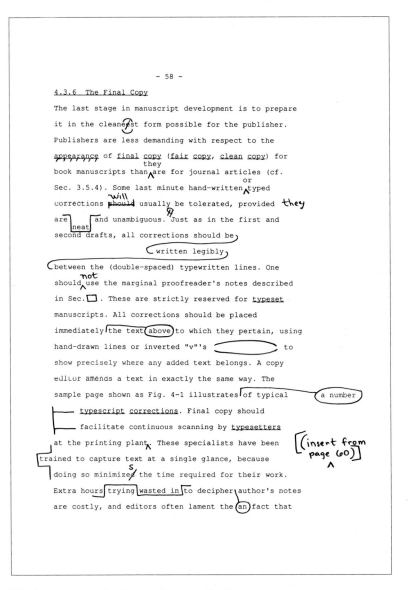

Fig. 3–1. An example of a page from an edited manuscript.

or perhaps suggestions of changes that might enhance its "professional" tone.[5] The next step is to introduce typesetting instructions. This can be a particularly arduous task in the case of a scientific document because of the frequent occurrence of equations and other complex symbolic expressions. The editor begins by addressing more routine matters, however: establishing a line length, typeface, and type size for the body text, setting a standard for paragraph indentation, etc. Words or phrases that require special highlighting must also be marked, with indications in each case of the *nature* of the proposed highlighting (italic, boldface, etc.). Certain parts of the text may also be singled out for setting in unusually small type, typically passages the editor feels are "less important" for the average reader. Other paragraphs may be marked for exceptionally wide margins, again as a way of ensuring that they will stand out from the body of the text.

As suggested above, the marking process can become rather complicated if the text contains a great many mathematical expressions. Each symbol present must be carefully scrutinized and equally careful marked to ensure that it will be processed correctly. For example, every character representing a *physical quantity* or a *variable* must be accurately identified (cf. Sec. 3.4), including ones that occur within the running text. Special attention must be devoted to any symbols that could prove ambiguous. A marginal note might be added to warn the typesetter that a particular character is in fact the Greek letter "nu", for example, since the *symbol* "ν" bears a disconcertingly close resemblance in many typefaces to the European letter "v"; similar advice would apply to the number "zero" ("0") in any context where it might be mistaken for the capital letter "O".

Normally all this typesetting information would be directed toward a professional typesetter, whose responsibility it would be to "retype" the entire manuscript as a first step in the printing process—but it could also be addressed to an author who has agreed to supply the publisher with formal camera-ready copy. Like an editor, the desktop publisher preparing to typeset an electronic manuscript should examine the document in a very systematic way in a search for specific characters that require special treatment. Markings should be introduced *by hand*, preferably in red or some other

---

[5] Publishing houses vary considerably in the extent to which they encourage this particular type of editing. It is obviously time-consuming (and costly), and it presupposes the availability of dedicated, talented, and highly trained editors, but it is also the key to maintaining consistently high publishing standards.

distinctive color, in a hardcopy version of the manuscript using an appropriate set of formatting marks (e.g., wavy underlines for italics, double-underlining for boldface, etc.). Changes should then be made in the corresponding computer files as a separate, subsequent step. (An experienced author–typesetter may well enter most of the symbols correctly from the outset.)

Following this general introduction to the problem we now turn our attention to those text elements that most often require special treatment, and to what that treatment might be.

3.3 Symbols Appropriately Set in Standard (Roman) Type

We begin our systematic discussion of the typography of science with the comforting message that some symbols are in fact "normal", in that they are always set in ordinary (roman) type. These include:

● *Numbers;*[6] e.g.,

> 12   0.765 12   123 078.01

However, a long-standing typesetter's rule holds that if reference is required in running text to a number between one and twelve—in the absence of larger numbers—it should be expressed in words, as in this sentence. Larger numbers, such as 13 or 250, and smaller numbers like 3 that may accompany them, are indicated symbolically.

● *Unit symbols*, such as[7]

> m   L   kg   s   A   K   mol   cd

---

[6] Numbers serving as *labels* for parts of illustrations represent a special case: these are traditionally italicized.

[7] Note that the "L" for "liter" is *capitalized* to avoid possible confusion with the numeral "1". Unit *symbols* like these are appropriate only in conjunction with numerical values. Notice also that these are regarded as true symbols and not abbreviations, so they are *not* followed by periods (with the exception of "in." for inch). The corresponding *words* should be used in references to the units alone ("… to each liter of solvent one should add …").

- Symbols for *proportions*, including

  ‰   ppm   ppb

- *Arithmetic signs and symbols*, as well as symbols from set theory and mathematical logic; e.g.:[8]

  $$+ \ - \ : \ \times \ > \ < \ \oplus \ \otimes \ \in \ \varnothing \ \cup \ \cap \ \exists \ \vee \ | \ \Rightarrow \ \Leftrightarrow$$

As noted in Section 2.11 the proportional typefaces available with modern computer systems normally include three representatives of the "dash" family. Only one of these—the medium-sized "en-dash"—is suitable for use as a minus sign. Unfortunately for mathematicians, the symbol evoked automatically by the "dash" key itself is always the shorter *hyphen*, which makes a very unsatisfactory substitute; e.g.:

  $$x - 4 \ \text{or} \ \text{Cl}^- \quad (not \ x\text{-}4 \ \text{or} \ \text{Cl}^\cdot)$$

- Symbols for *operators*, such as

  $$\Delta \ \ d \ \ \partial \ \ D \ \ \delta \ \ \nabla$$

- *Functions* of the exponential, logarithmic, or trigonometric type; e.g.:[9]

  $$\exp \ \ \log \ \ \ln \ \ \sin \ \ \tan \ \ \sinh \ \ \tanh$$

as well as such analogous special functions as the hermitic and Laguerre polynomials, the gamma function, and the Riemann zeta function:

  $$H_n(x) \quad L_n(x) \quad \Gamma(x) \quad \varsigma(x)$$

- π and e (the basis for the set of natural logarithms)

- i (the unit for imaginary numbers), as well as the functions Re (real portion) and Im (imaginary portion); e.g.:

  $$z = a + ib \quad \text{Re}(z) \quad \text{Im}(z)$$

Apart from roman, the single most important type style in scientific text is *italic*, as will be apparent from the section that follows.

---

[8] For more extensive lists of such symbols see *The Chicago Manual of Style* (University of Chicago Press, 1988), Beyer (1987), or Vol. II (pp. 1522–1528) of Iyanaga and Kawada (1977).

[9] For a more comprehensive list see Swanson (1986).

3.4 Symbols That Demand Special Typefaces: Variables, Functions, Quantities, Vectors, Tensors, and Natural Constants

Symbols used to represent *physical quantities* or *mathematical variables* are almost always single letters from the Latin or Greek alphabets. Only in rare cases is a standard quantity symbol composed of more than one letter (an example is *Re*, the Reynolds number, with the dimension 1). To prevent confusion, complex symbols of the latter type are often set in parentheses, as in (*Re*).

Quantity and variable symbols are occasionally supplemented with *subscripts* as a convenient way of adding specificity, as in DH_{vap}, the enthalpy of vaporization. Similarly, the *dependency* of some quantity on a particular variable (such as time) might be indicated by a subscript (e.g., v_t for velocity as a function of time; for reasons discussed in Sec. 3.6 *both* symbols in this case have been set in italic type). Alternatively, the symbol for such a variable might be set in parentheses immediately after the symbol for the quantity of primary interest, as in $v(t)$. In rare cases specificity is conferred by a *superscript* rather than a subscript (e.g., $H°$ for *standard* enthalpy[10]), or by some other special mark such as an overbar ($\bar{c}$), a caret ($\hat{L}$), or a tilde ($\tilde{v}$). Typographic considerations applicable specifically to subscripts and superscripts are discussed in Section 3.6.

All the various symbols with which we are presently concerned require the availability of special typefaces, a concession designed to assist the reader in interpreting their significance. The most important typographic rules specifying the use of non-standard type may be summarized as follows:

● Symbols for *mathematical variables* are always set in italic type; e.g.:

$$x, y, z, f, A, B, C$$

---

[10] IUPAC also sanctions use of the superscript "⊕" to indicate a standard value, although this notation has not been universally adopted. Many chemists have also been reluctant to implement the IUPAC recommendation that the standard enthalpy of vaporization, for example, should be expressed as $\Delta_{vap}H^{\ominus}$, preferring instead the more traditional placement of a subscript like "vap" on the quantity symbol rather than on the Δ; cf. IUPAC (1988), pp. 44 ff.

- Symbols for *general functions* (as distinct from the special functions treated in Sec. 3.3) must also be set in italic type; e.g.:

$$f(x) = u(x)/v(x)$$

- In a similar way, italic type is required for any symbol that represents a *physical quantity*; e.g.:

$$l \text{ (length)}, \ t \text{ (time)}, \ T \text{ (thermodynamic temperature)}$$

- The same rule applies to *natural constants*, such as N_A (Avogadro's constant), h (Planck's constant) or R (the universal gas constant).

In other words, these are treated typographically as if they were variables, despite their presumed "constancy". The explanation for this seeming anomaly is to be found in the close relationship between constants of this type and various physical quantities. Note, however, that the *subscript* in the case of N_A is *not* italicized, because it plays the part of an identifier (cf. Sec. 3.6).

- Symbols representing *vectors*[11] or *directed physical quantities* should be set in *bold italic* type; e.g.:

$$\boldsymbol{a} = a_1\boldsymbol{e}_1 + a_2\boldsymbol{e}_2 \qquad \boldsymbol{F} \text{ (force)} \qquad \boldsymbol{E} \text{ (electric field strength)}$$

- *Tensor* symbols are the subject of a rather unusual rule: they are to be set *bold roman*, but in a *sans-serif* font like Helvetica; e.g.:

A B C

3.5 Spacing Conventions

Normally a typesetter would be free to exercise a considerable amount of judgment with respect to optimum character spacing, particularly in the interest of improving readability, but certain situations involving mathematical expressions *demand* the presence (or absence) of blank space, again based on international convention.

---

[11] Vectors were once commonly represented by gothic (better: *German*) characters (i.e., 𝖆, 𝖋, 𝕰), or by quantity symbols surmounted by arrows (i.e., $\vec{a}$, $\vec{F}$, $\vec{E}$), but these conventions are now discouraged.

● Numbers consisting of more than four adjacent digits should be grouped into *spaced triads* (sets of three) starting from the decimal point; i.e.:

> 17 315   7 215.01   0.000 000 3   3 435 213.010 45

Four-digit numbers are normally set as a block, as in "3725", except in tables that also contain grouped triads, where consistent spacing should be maintained throughout.

● In the symbolic representation of a *quantity*, the *unit* symbol must be separated from the corresponding *numerical value* by a single space:

> 3 m   13 °C   280.05 K   17.5 kg   12 mol/L

Note that the *entire* symbol for "degrees Celsius", °C, is set apart from the number (i.e., "13 °C", *not* "13°C" or "13° C"). On the other hand, the degree symbol for an *angle* is set flush with the corresponding numerical value, a convention that also applies to the symbols for angular minutes and seconds; e.g.:

> 180°   12.3′   51° 12′ 35.5″

● No space (or only a thin space) should be left between a number and an associated *proportion* symbol [percent (%), "per-thousand" (‰)], although a full space is of course required before a proportional "unit" like ppm or ppb (parts-per-million or parts-per-billion[12]):

> 12.4%   0.1‰   20 ppb

Space is also omitted between a number and a symbol it is understood to multiply (e.g., $2x$).

● A full space separates *pairs* of unit symbols understood to be related in a multiplicative sense, as in:

> $0.7 \text{ g cm}^{-1} \text{ L}^{-1}$

● Single spaces (or thin spaces) should be left *before and after* the symbols for addition, subtraction, multiplication, division (÷), and equality:

> $z = a + b$ 3.81×10^{-6} $1.041 \cdot 10^{12}$

---

[12] This is a convenient place to note that the term "billion" is best avoided, especially in scientific text, since the meaning it conveys in the U. S. (10^9) is three orders of magnitude *smaller* than the value associated with a billion in many other countries.

- The same applies to integral, summation, and product-formation symbols:

$$3\int_a^b f(x)\,dx \qquad \sum_{i=1}^{n} x_i \qquad \prod_{i=1}^{5} (8-i)^{i^2}$$

However, when "+" or "–" is used in an adjectival sense it should be set quite close to the symbol to be modified (e.g., –3 °C).

- A single space is required both before and after a symbol for one of the mathematical *functions* (such as the logarithmic function, or one of the trigonometric functions), *except* when followed by an expression in parentheses:

$$\sin x \qquad 4 \tan 3y \qquad \cos(3x+y)$$

Special care should be exercised in the introduction of discretionary spaces, because sometimes the presence or absence of space will significantly alter the apparent meaning of an expression. This is particularly true with expressions containing function symbols. Consider, for example:

$$a \sin \omega t\, e^x$$

The spacing shown suggests that the terms ωt and e^x are independent entities, only the first of which is subject to the function "sin". A very different interpretation would be associated with the expression

$$a \sin \omega t e^x$$

Potential misunderstandings of this type are most easily avoided by introducing parentheses:

$$a \sin(\omega t)\, e^x \quad \text{or} \quad (a \sin \omega t)\, e^x$$

3.6 Subscripts and Superscripts

Subscripts and superscripts *(indices)*, including *exponents*, should always be set in type smaller than that used for the principal characters to which they are attached, typically with a size reduction of 35–40%. Thus, indices used in conjunction with 12-point type might be as small as 7 or (preferably) 8 points, although even 9 points would not be considered excessive.[13]

Indices should always be set directly adjacent to the characters to which they apply. If a particular symbol requires both a subscript *and* a superscript, then ideally the two indices would be set so that one falls directly below the other; e.g.:

$$K_S^2 \quad \text{or} \quad v_{\max}^2$$

If this arrangement is not feasible with the available word-processing or page-layout software, permissible alternatives include:

$$K_S{}^2 \quad \text{and} \quad (K_S)^2 \quad \text{or} \quad v_{\max}{}^2 \quad \text{and} \quad (v_{\max})^2$$

The same principle applies in the case of chemical formulas involving charged species:

$$NH_4^+, SO_4^{2-} ; \text{if necessary, } NH_4{}^+, SO_4{}^{2-} \quad \text{or} \quad (NH_4)^+, (SO_4)^{2-}$$

Note that the first alternative proposed places the subscript *before* the superscript, but a *prime* symbol (e.g., K'_2) is always set next to the symbol to which it applies.

Any index that represents a physical quantity should be set in italic type. Thus, the symbol c_p refers to the specific heat *(c)* at a constant pressure *p*, so both the c and the p are italicized. On the other hand, g_n, which stands for the normal acceleration due to gravity, is typeset with a roman subscript, since the "n" in this case is simply an abbreviation for "normal" (see Table 3–1 for a list of common indices and their meanings).

---

[13] Most word-processing and page-makeup programs apply a preset (default) reduction factor automatically to indices, but a full-featured program should also provide facilities for *adjusting* this parameter—as well as the extent of vertical displacement—to suit the user's needs.

Table 3–1. Frequently encountered indices and their significance (taken in part from IUPAC, 1988).

Index	Significance	Index	Significance
Subscripts:		p	polar
ads	adsorption	r	radial, reaction (in general)
at	atomization	rel	relative
ax	axial	red	reduced
c	combustion	sol	solution
dil	dilution	sub	sublimation
eq	equivalent, equatorial	vap	vaporization
e	electrical		
exp	experimental	*Superscripts*:	
eff	effective	⊖, o	standard
f	formation	*	pure substance
fus	fusion (melting)	λ	infinite dilution
max	maximum	id	ideal
min	minimum	‡	activated complex,
n	normal component,		transition state
	normal condition	E	excess quantity

Multiple indices of the same type that apply simultaneously to a single quantity or variable should be set adjacent to each other and on a common level, but separated by a comma or a space, or distinguished by parentheses; e.g.:

$$V_{T,p} \qquad V_{Tp} \qquad V_{T(p)}$$

If an index itself bears an index, the result is a three-level expression, and the third-level index must be set slightly below (or above) the second-level index and in correspondingly smaller type, as in e^{x^2}. Expressions of this sort are often difficult to read and interpret, however, so they should be avoided whenever possible. Consider, for example, the expression: $x_{n_{max}}$. In this case no ambiguity is likely to result from use of one of the more legible presentations:

$$x_{n,\,max} \quad \text{or} \quad x_{n\ max}$$

That is, the "n" and "max" might be treated as though they were equivalent, and simply separated by a comma, or the subordinate status of the "max" could be implied by the use of somewhat smaller type.

3.7 *Typesetting Guidelines for Physical Equations and Formulas*

The following guidelines apply quite generally to typeset physical or mathematical equations and formulas (cf. Swanson, 1988):

- If possible, equations and formulas should be isolated from surrounding text (*displayed*, cf. Sec. 2.7), despite the fact that this practice increases the length (and therefore the cost) of a document. *At least* one-half or one full line of blank space (leading) should be inserted above and below such an expression, which can also be indented somewhat relative to the surrounding text. Whatever display standards are adopted, they should be maintained consistently throughout a given document.

In a manuscript containing a large number of equations, those that are actually referred to in the text should be numbered—and then referred to by their numbers. The *equation number* is set on the same line as the equation itself, usually in parentheses and flush against the right margin; e.g.:

$$q_c = (q_a + 3q_b)^2 \qquad\qquad (47)$$

Correct placement of equation numbers is achieved most easily with the help of a word-processor's *right tab* function.

- Very short equations (e.g., $x = 43\,z$, or $h = 24.5$ mm) can be incorporated directly into running text, but this is accompanied by the risk of serious spacing problems in fully justified text, since there is no acceptable way to divide an expression like $\exp[-E_a/(RT)]$ in the unfortunate event that it happens to fall near the end of a line. Similarly, a line break should never be allowed to separate a numerical value from its associated unit symbol(s).

- Some freestanding equations are too long to fit in the space available on a single line. When this is the case the equation should be interrupted immediately *after* an operation symbol, and the symbol itself should be repeated at the beginning of the continuation line; e.g.:

$$c^4(H_3O^+) + c^3(H_3O^+) \cdot K_{S1} + c^2(H_3O^+) \cdot (K_{S1}K_{S2} - K_w - CK_{S1}) +$$
$$+ c(H_3O^+) \cdot (CK_{S1}K_{S2} - K_{S1}K_w) - K_{S1}K_{S2}K_w = 0$$

The best place to divide a long expression (e.g., $a_x = b + q + \ldots = \ldots$) is at an "equals" sign. This produces a *set* of expressions, which can then be so arranged that corresponding equality symbols are all cleanly aligned (by careful placement of appropriate tab markers!);[14] e.g.:

$$
\begin{aligned}
p_0 &= k_1 h^2 N^{5/3} m^{-1} \\
&= k_2 N \varepsilon_{\mathrm{F}} \\
&= k_2 U_0 V^{-1}
\end{aligned}
$$

- Several alternatives are available for expressing the operation of multiplication:

$$ ab \qquad a \cdot b \qquad a \times b $$

Note that the multiplication symbol in the third example is a true "cross" (*not* simply the letter "x"), taken in this case from the font Symbol, and that it appears to be set slightly *above* the baseline.

- When representing a product of *vectors* it is important to distinguish clearly between a *scalar product* ("dot product")

$$ \boldsymbol{ab} \text{ or } \boldsymbol{a} \cdot \boldsymbol{b} $$

and a *vector product* ("cross product"). A product of the latter type is always expressed by a cross symbol:

$$
\boldsymbol{a} \times \boldsymbol{b} =
\begin{vmatrix}
\boldsymbol{e}_1 & \boldsymbol{e}_2 & \boldsymbol{e}_3 \\
a_1 & a_2 & a_3 \\
b_1 & b_2 & b_3
\end{vmatrix}
$$

- The operation of division can also be represented in several ways:

$$ \frac{a}{b} \qquad a \div b \qquad a : b \qquad a/b \qquad a\, b^{-1} $$

The first alternative, "stacked" notation, is the one most easily interpreted at a glance, a point especially worth considering in the case of complicated expressions.

---

[14] For further advice on breaking long mathematical expressions see Swanson (1988).

Compare, for example:

$$z = \frac{\dfrac{1}{xy} + \dfrac{1}{\ln x}}{\dfrac{x}{x+y} + \dfrac{y}{x-y}}$$

$$z = [1 : (xy) + 1 : (\ln x)] : [x : (x + y) + y : (x - y)]$$

$$z = [1/(xy) + 1/(\ln x)] \,/\, [x/(x + y) + y/(x - y)]$$

$$z = [x^{-1}y^{-1} + (\ln x)^{-1}][x(x + y)^{-1} + y(x - y)^{-1}]^{-1}$$

Unfortunately, the preferred notation is also the most difficult to typeset, especially if one lacks access to software created especially for the purpose.[15]

Note that use of a "slash" symbol[16] to represent division requires careful attention to *grouping symbols* (see below) in order to prevent ambiguity. Thus, the expressions

$$a/b/c \quad \text{and} \quad \ln x/3$$

are both ambiguous, and open to the following alternative interpretations:

$$a/(b/c) = (ac)/b \quad \text{or} \quad (a/b)/c = a/(bc)$$
$$\ln (x/3) \quad \text{or} \quad (\ln x)/3$$

Slash notation is usually preferred for fractional exponents.

● The symbol for an integral, summation, or product should be of a size commensurate with the expression to which it applies; e.g.:

$$\sum_{i=0}^{\infty} f(x_i) \qquad \sum_{i=0}^{\infty} \frac{f(x_i)}{g(x_i)}$$

---

[15] Version 5.0 of WORD for the MACINTOSH includes special auxiliary software for creating complex mathematical expressions. Other programs for the typesetting of mathematical expressions are discussed in Sec. 9.7.

[16] Technically, the symbol shown in this example should be referred to as a "virgule". Its effectiveness in the example above has been improved slightly by making it slightly larger than the surrounding characters, and by a modest downward displacement relative to the baseline. Many computer type fonts also provide access to an alternative, less steeply sloped slash known as a "solidus" (/, Macintosh key combination "shift-option-1") designed specifically for typesetting in-line fractions, but effective use of this symbol mandates special size and height adjustments for the terms constituting the numerator and denominator (e.g., $^2/_3$).

In the same way, parentheses and related grouping symbols *(fences)* should be made large enough so that they actually "contain" their contents. The preferred sequence for the *nesting* of such symbols is usually:

$$\{ \, [\, (\,) \,] \, \}$$

3.8 Chemical Formulas

Chemical formulas typically convey information more efficiently than "official" compound names accepted by IUPAC. Even an experienced chemist would be unlikely to recognize in the cumbersome name

pentacyclo[4.2.0.0^{2,5}.0^{3,8}.0^{4,7}]octane

the presence of eight carbon atoms joined in the symmetrical pattern apparent immediately from the compound's structural formula:

In this case, of course, the trivial name "cubane" is equally revealing, but most trivial names provide the non-specialist with little more structural insight than the corresponding IUPAC names.

A clear distinction should be made at the outset of our considerations between *molecular formulas* and *structural formulas*. A molecular formula, such as $C_6H_{16}F_3OS_2$, is nothing more than a list of elements, with each element symbol subscripted to indicate the number of atoms of that type present in a molecule of the substance in question. Traditionally, the "organic" elements carbon and hydrogen are listed first in a molecular formula, followed by other elements arranged in alphabetical order.[17] Formulas of this type

---

[17] More information on nomenclature and the notation applicable to chemical compounds is available from many sources, ranging from the extensive sets of authoritative rules developed by IUPAC and *Chemical Abstracts* to brief summaries in introductory text books. A relatively concise overview of the subject is available from Ebel, Bliefert, and Russey (1987).

can usually be incorporated without difficulty into running text, although it may occasionally be necessary to revise a sentence to prevent a long formula from coming too near the end of a line (since a line break within such a formula is prohibited). Care should also be exercised to ensure that subscripts in a formula do not distort the line spacing of the text (see also Sec. 9.2).

In addition to subscripts that specify numbers of atoms, indices of other types are sometimes associated with element symbols as well; e.g.:

upper left	mass number (number of nucleons)
lower left	atomic number (number of protons)
upper right	electric charge symbol (+ or –), *preceded* as necessary by a number to indicate the *extent* of the charge on that particular atom; this position is also used for an asterisk or some other indication of an excited atomic state

e.g.: $^{32}PCl_5$, $^{15}NH_4^+$

A few words are perhaps in order here with respect to the typography of *names* of chemical compounds. For instance, certain name fragments should always be set in italic type. This applies specifically to *prefixes* (including Greek letters) used for distinguishing particular isomers or stereoisomers, which are then separated from the names themselves by hyphens. The most important prefixes of this type are:[18]

> *sec* or *s* (secondary), *tert* or *t* (tertiary), *n* (normal), *o* (ortho), *m* (meta), *p* (para), *cis*, *trans*, *syn*, *anti*, *meso*, *epi*, *cyclo*, *R*, *S*, *E*, and *Z*

For example:

> *cis*-2-butene; (1*S*,2*S*)-1,2-dimethylcyclopropane

The same rule applies to the letters (*a*, *b*, *H*, etc.) that are sometimes used as *locants* within a compound name.

---

[18] The isomeric designations D and L represent exceptions to the rule. Actually, the use of these (obsolete) symbols is now discouraged, but *if* they appear they should always be set in roman type, and as *small capitals*; e.g.:

L-isoleucine, D-Ribose

Atom symbols that constitute part of the name of a compound are also italicized; e.g.:

O-methylhydroxylamine, N,N-dimethylhydrazine

In contrast to a molecular formula, a *structural formula* is expected to communicate not only the number of atoms of each kind present, but also at least some sense of the way in which specific atoms are interconnected. Like ordinary molecular formulas, relatively simple structural formulas, especially linear or *condensed* formulas, can be incorporated directly into running text, although some care is required to ensure clarity, and problems may again be encountered with awkward line breaks, especially in fully justified text. A *single bond* in such a formula should be symbolized not by a hyphen, but by an "en-dash" (cf. Sec. 2.11), as in $H_3C–Br$. Double bonds can be expressed reasonably well in most typefaces with an "equals" sign (e.g., $H_2C=O$), but specifying a triple bond (e.g., $N\equiv N$) requires access to an appropriate set of special symbols, such as the general-purpose font with the *name* Symbol (cf. also Sec. 3.9). Typical examples of linear structural formulas include:

$H_2N–CH_2–CH=CH–CH=CCl–CHClF$
$H_3C–C_6H_4–SO_2–NHOH$

Note that the first example is still ambiguous with respect to *cis,trans* stereochemistry about the double bonds, and the second fails to specify relative positions for the substituents on the benzene (C_6H_4) ring.

The typography becomes much more problematic if one wishes to introduce a spatial representation of an inorganic crystal structure, for example, or a depiction of the precise arrangement of atoms in a complex organic molecule. One way of approaching the problem is to take advantage of graphics software (cf. Chap. 11) and simply sketch the appropriate structure, just as one might on paper. Pixel-based "paint" software should be avoided, however, because the high resolution appropriate to a laser printing device is very difficult to achieve in this way (and impossible with some programs), and images that result are extremely memory-intensive (cf. Sec. 11.2). Laser-quality representations are best prepared with an *object-oriented* graphics program (cf. Sec. 11.3). For example, Fig. 3–2, which depicts the three-dimensional structure of a silicate, was created using the program CRICKETDRAW. It consists of relatively simple arrays of straight lines and plain or shaded circles. Illusions of depth were achieved by specifying lines of various widths or gray patterns with differing intensities.

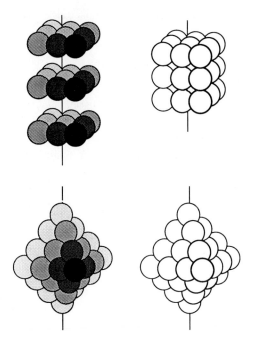

Fig. 3–2. Graphic depictions of silicate structures, represented in terms of spheres, and prepared with the object-oriented graphics program CRICKETDRAW.

In most cases chemical structure diagrams can be prepared most easily with special software designed specifically for that purpose. Figure 3–3 illustrates several structures created with the MACINTOSH program CHEMIN-TOSH, which is discussed in some detail in Sec. 11.6.[19]

Whenever graphics programs are used to prepare subsidiary elements for a text manuscript (chemical formulas, mathematical equations, etc.), some attention must be directed to the most appropriate *file format* for saving the results. Graphics programs typically support several different file formats (PICT, TIFF, etc.), some of which may be incompatible with word-processing or page-layout files containing the body of the text (for more information see Sections 8.4 and 11.6).

---

[19] The developers of CHEMINTOSH (SoftShell International) offer an analogous program for IBM-compatibles called CHEMWINDOWS. Files prepared with either program can be interpreted by the other.

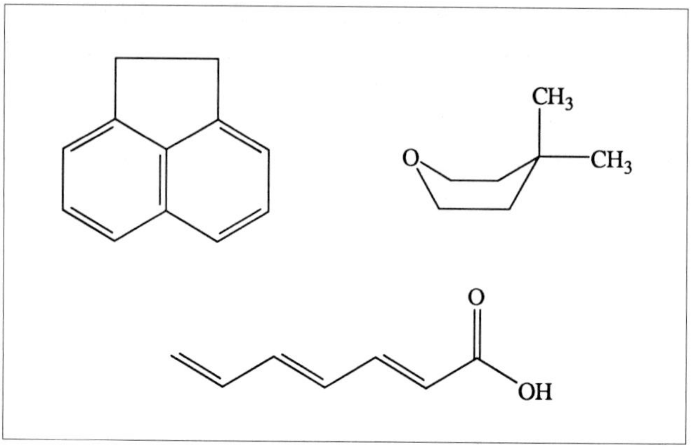

Fig. 3–3. Typical chemical structures prepared with the aid of CHEMINTOSH.

3.9 Special Symbols

Scientists regularly confront the need to incorporate into their documents symbols not provided for by the ASCII code, and therefore not represented in most computer type fonts; e.g.:[20]

$$\wp\ \partial\ \triangle\ \angle\ \Im\ \uparrow\ \aleph\ \pm\ \infty\ \neq\ \lrcorner\ \vee\ \leftarrow\ \cap\ \otimes\ \approx\ \varnothing\ \Re$$
$$\oplus\ \times\ \Leftarrow\ \Uparrow\ \rightarrow\ \in\ \Downarrow\ \Leftrightarrow\ \cup\ \downarrow\ \div\ \Rightarrow\ \cong\ \exists\ \notin\ \bot\ \supset$$

The particular symbols shown here do not constitute a problem: all of them are included in the versatile font *Symbol*, which is provided as a standard feature with most laser printers (cf. Fig. 2–10). Symbol is only one example of what is sometimes referred to as a "pi font", a computer character set in which ASCII code values have been arbitrarily reassigned by the font designer to an alternative set of characters. Accessing one of these characters from the computer keyboard is simply a matter of selecting the appro-

---

[20] For more information on symbols commonly encountered in physics see the article "Symbols, Units, and Nomenclature" by E. R. Cohen in Lerner and Trigg (1991), pp. 1217–1232, as well as Parker (1989). Sources of mathematical symbols are listed in footnote 8 on p. 94.

priate font and then consulting a table of equivalencies to ascertain the appropriate keystroke combination. Other (downloadable) fonts of this type include Lucida Math, Mathematical Pi, and Universal Greek with Math Pi, all available from Adobe Systems.[21]

Special symbols play an important role in typesetting for a wide variety of scientific fields (cf. Boshard, 1980), including:

- astronomy
- electronics
- automation control engineering
- data processing
- botany
- metallography
- meteorology

Few professional printers have ready access to all the symbols of potential interest, but special sources have been developed to meet at least some of the needs of science-oriented desktop publishers.[22] Nevertheless, one should also be prepared on occasion to construct a custom symbol within the context of an appropriate graphics program. Symbols created in this way can easily be imported into the text of any document—as often as required, and at any desired scale—in the form of "in-line" graphics (cf. Sec. 9.4.2). In some cases it may even be worthwhile to consider using a program such as FONTOGRAPHER (Altsys) to create a custom font containing important special characters, thereby rendering them instantly and permanently accessible from the keyboard.

---

[21] It should be noted in passing that at least some standard text fonts also provide access to a surprising range of symbols, which are assigned to special (non-intuitive!) keystroke combinations.

[22] For example, Nisus Software distributes a set of special downloadable laser pi fonts called TECHFONTS, which includes both SCIFONTS (scientific characters) and ELECTROFONTS (analog and digital circuit symbols), and Linotype-Hell offers a set of seven fonts called CHEMICAL PI.

Part II
Hardware

4 Computers

4.1 Text-Editing Systems in General

Chapter 1 dealt briefly with some of the hardware and software developments that have now made it possible for almost anyone with a personal computer and sufficient enthusiasm to prepare true camera-ready copy. In this chapter we begin to look a bit more closely at a few of the specific options with respect to hardware.

A basic *writer's work station (text-editing facility* or *word processor)* consists of a keyboard, a screen, a central processing unit with an adequate amount of memory, and a printer. To this must of course be added an appropriate set of software, as discussed in Part III. More specifically, the hardware prerequisites for convenient and flexible creation and manipulation of scientific text and illustrations can be summarized as:

- a personal computer with its associated display device and keyboard (Sections 4.2, 4.3 and 6.3),
- a set of disk drives, including both a "hard disk" for bulk storage (Sec. 5.3) and one or more drives for diskettes (or "floppies"; Sec. 5.2), the most common portable media,
- a laser printer, ideally complemented by access to a laser typesetter ("imagesetter"; cf. Chap. 7), and
- a scanner (Chap. 8).

Software needs include:

- a good word-processing program (Chap. 9),
- a selection of graphics programs for the preparation of charts and illustrations (Chap. 11),
- a page-makeup program for combining and arranging information initially saved as text and graphics files (Chap. 10),

- various pieces of special software to facilitate the typesetting of mathematical equations and chemical formulas, for example, or to assist in transforming raw numerical data into graphic form (Chap. 12), and, in some cases,
- programs that allow one to edit scanned illustrations, or interpret scanned text (Sections 8.2 and 8.5).

An optimum system also reflects the following general characteristics:

- ease of use with respect to all its components
- minimal training requirements
- support for a wide range of file and disk formats
- a high-resolution video display (monitor)
- adequate provision for future upgrading

In reaching a decision with respect to a particular system one should be sure to take into consideration not only the characteristics of the individual components but above all the extent to which the various parts interact smoothly with one another. No system can ever be more powerful than its weakest link, and lack of full compatibility is a distressingly frequent complaint with some systems.

A great many computer systems offer the potential for meeting these basic tests, but one factor deserves to be singled out for special mention: *convenience*. Nothing is more frustrating for the serious writer than being diverted from creative work by a host of computer "bugs", or the feeling that a computer expert would be required to carry out some seemingly trivial task. From the standpoint of the writer, the computer is, after all, only a *tool*, and tools are supposed to make life simpler, not complicate it. This was the basic philosophical position adopted by executives at Apple Computer in Cupertino, California, in the early 1980s as they guided the development of one of the first true desktop publishing packages, with the MACINTOSH computer at its heart. Apple's MACINTOSH operating system (cf. Sec. 4.4) was widely praised for being extraordinarily "user-friendly", and many who worked with it recognized its potential for influencing the evolution of desktop publishing. Indeed, the advent of the MACINTOSH can fairly be regarded as a decisive event in the "desktop-publishing revolution".[1]

Probably the most powerful argument in favor of the MACINTOSH desktop-publishing system today, ten years later, is the remarkable degree to which it continues to reflect outstanding compatibility with respect to both software and hardware despite explosive growth in the scope and variety of available programs and accessories—designed, maintained, and distributed by a host of companies, both large and small. The authors of the present book have closely followed the evolution of MACINTOSH writing and publishing tools virtually since their inception, which helps explain why so many of the examples we cite are drawn from this reservoir.

Much of the attention the MACINTOSH received in the early years can be attributed to the new role assigned to the display screen, and more specifically to Apple's approach to communication between the computer and the user. Fresh thought and innovative programming techniques were combined in an effort to reshape this communication to reflect more clearly the needs of the *user*. Traditional computer jargon was cast aside, as was the artificial rigidity imposed by the traditional "command line" mode of interaction (cf. Sec. 4.3). One consequence was that many computer aficionados haughtily dismissed the new "toy computer", but the inherent wisdom in Apple's bold departure has been more than confirmed by the recent rapid swing within the IBM-compatible (MS-DOS) world toward an analogous operating environment, MICROSOFT WINDOWS. "MACINTOSH-like" DTP software has become an important component of the WINDOWS market, and many of the most important DTP programs—including PAGEMAKER, QUARKXPRESS, WORD,

---

[1]   Baeseler and Heck (1987) were among the first Europeans to acknowledge the importance of the MACINTOSH phenomenon, reporting that "The Apple MACINTOSH, together with a LASERWRITER, a desktop-publishing program such as PAGEMAKER or READY,SET,GO!, and a few subsidiary programs today constitutes the most highly developed among the commercial [DTP] systems, one that meets the demands even of professionals. Desktop publishing is only beginning to see the light of day in the IBM world, but the Macintosh can already point to everyday practical applications." An article in the respected German business journal *Handelsblatt* (3 February 1988) went so far as to assert that: "... the abbreviation 'DTP' has evolved into the keyword of an industry and a synonym for one manufacturer (Apple), which has advanced from a pioneer in the business to the market leader." Finally, M. Breuer in an article entitled "Layout and Page Makeup by Hand" [*Computer Persönlich*, special issue *Textverarbeitung*, **1987**, 152] concluded that "For low-investment desktop publishing purposes there in fact exist today only two systems, the Apple MACINTOSH and the IBM-PC (including compatibles). Now as in the past the major force in the marketplace is the MACINTOSH, which together with the Apple LASERWRITER constitutes the DTP system with the fewest 'rough edges'".

ADOBE ILLUSTRATOR, PHOTOSHOP, CANVAS, and many others—are now available in similar (and fully compatible) versions for both environments.[2]

As stated previously, our primary intent here is to provide insight into the *techniques* applicable to contemporary desktop publishing, most of which are now largely independent of any particular hardware platform.[3] Nevertheless, since much of the historical development of these techniques is closely linked to the development of the MACINTOSH computer family, a brief description of the latter seems a good place to begin.

4.2 The MACINTOSH Computer Family

The immediate precursor (1983) to today's MACINTOSH was an ill-starred computer with the nickname LISA, an acronym for "Local Integrated Software Architecture". LISA, later renamed the MACINTOSH XL, was a "... personal computer (developed by Apple, Inc.) based on an operating system whose user access evolved out of the applications sector in the form of symbols and pictograms. No longer was program management constrained by a keyboard; instead, much of it occurred by way of a 'mouse'. The technology involved can be regarded as truly trail-blazing from the standpoint of simple, uncomplicated, and user-friendly computer operation" (Schulze, 1988).[4]

Based on its early experience with LISA (and knowledge gained from a few serious engineering and marketing mistakes!), Apple next introduced the prototype MACINTOSH computer, conceived from the outset as a stand-alone system especially adapted to convenient editing of text and graphics—and

---

[2] A helpful introduction to challenges facing those who deal with the two computer environments simultaneously has been provided by G. Gruman in the article "Working in 2 Worlds: A Guide to Using the Same Applications in Macs and Windows PCs" (*Macworld*, December 1993, pp. 111–117).

[3] In fact, the "two worlds" of personal computing are drawing ever closer together, as underscored by Apple's recent release (March, 1994) of the first "PowerPCs", precursors to a line of very fast computers capable of working simultaneously in both the MACINTOSH and WINDOWS program environments (cf. Sec. 4.3).

[4] A "mouse" is an innovative input device—discussed in detail in Sec. 4.4—associated most closely with the MACINTOSH, but also utilized now by many MS-DOS applications.

completely divorced from the market-dominating world of MS-DOS personal computers. All the hardware for the first MACINTOSH was developed by Apple itself, and the corresponding software was applicable *only* within the MACINTOSH environment. Moreover, all applications programs were carefully crafted to be fully compatible with one another.[5] This early insistence on full compatibility has been rigorously maintained even though the components of the system have undergone considerable evolution. As a result, illustrations prepared with essentially any MACINTOSH graphics program can easily be incorporated into text files derived from almost any MACINTOSH word-processing program, or any type of page-makeup files.[6] Frustrations commonly encountered with other systems as a consequence of printer or software incompatibility are almost unknown to MACINTOSH users.

Early critics of the MACINTOSH were justified in voicing serious concerns about a lack of sophisticated software alternatives, but steadily growing popularity soon resulted in a veritable flood of new programs; indeed, the problem for MACINTOSH users today is one of sifting through a jungle of competing claims in search of the *best* software for a particular purpose. Another major complaint long directed at the MACINTOSH was an unreasonable pricing structure that in effect placed too high a premium on convenience, but this situation has also changed dramatically in recent months, and there is no longer any sound economic reason for dismissing the MACINTOSH advantages.[7]

Originally (1984) there was only a single version of the MACINTOSH computer, a compact all-in-one unit with a built-in nine-inch screen and what seemed at the time a generous 128 KBytes of memory. This first model was effectively replaced only nine months later by the "fat Mac" with a memory

---

[5] Günder (1988) writes: "The Macintosh is unprecedented with respect to both hardware and software in the unique way it facilitates interaction between virtually all the available programs."
[6] In fairness it should be acknowledged that a file conversion step is sometimes required prior to introducing certain types of graphics into some word-processor files, but this rarely presents serious problems (for more information see Sec. 8.4).
[7] One important factor in recent price adjustments is a sharp decline in the cost of both memory and laser-printer technology, but another is the inevitable market impact of competitive graphic interfaces, especially the WINDOWS environment for IBM-compatibles. Moreover, powerful, user-friendly desktop-publishing programs like PAGEMAKER eventually became available for alternative systems, further stiffening the competition. Low-cost MACINTOSH "clones" may exert a significant influence on future market developments.

Table 4–1. Specifications applicable to several of the MACINTOSH computer models available in late 1993

Features	PERFORMA 410	PERFORMA550/ LC 520	QUADRA605/ PERFORMA75/ LC 475
Processor	16 MHz 68030	33 MHz 68030	25 MHz 68040
RAM/hard-drive capacity	4/80 MBytes	5/160 MBytes	4/80 MBytes
Approximate retail price (with monitor)	$1000	$1300	$1600

Features	QUADRA 650	QUADRA 840AV	POWERBOOK 165
Processor	33 MHz 68040	40 MHz 68040	25 MHz 68030
RAM/hard-drive capacity	8/230 MBytes	8/230 MBytes	4/80 MBytes
Maximum RAM	132 MBytes	128 MBytes	8 MBytes
Approximate retail price (with monitor)	$2500	$4500	$2000

of 512 KBytes, to be succeeded in turn by the MACINTOSH PLUS and the SE, all of which maintained a common physical profile. The pattern of uniform appearance was first broken in 1987 with release of the MACINTOSH II, which featured a separate central processing unit and a large display screen. In recent years the MACINTOSH product line has branched out in a bewildering number of directions, ranging now from the convenient and portable POWERBOOK models to the powerful and diverse QUADRA and POWERPC series.[8] Table 4–1 summarizes key features and characteristics of several advanced members of the MACINTOSH computer family.[9]

Despite the potential confusion engendered by this diversity, it remains true that almost any MACINTOSH computer supported by a PostScript-compatible laser printer (cf. Sec. 7.3) can be regarded as an effective platform

---

[8] For a more extensive review of the evolution of the MACINTOSH see the article "Macintosh Innovations" by G. Gruman and J. Heid (*Macworld*, February 1994, pp. 86–98).

[9] Long feature articles in the December 1993 issues of both *MacUser* and *Macworld* provide more extensive data and thoughtful qualitative comparisons.

for routine work with both text and graphics. Realistically, however, one should probably consider also acquiring some type of auxiliary data-storage device (either internal or external, cf. Chapter 5), both for routine backup purposes and because nearly everyone quickly manages to fill whatever storage space is initially available—with a multitude of text documents, miscellaneous pieces of software, alternative fonts, and unexpectedly large graphics files.

Perhaps the most important technical consideration with respect to the computer itself, at least from the standpoint of operating convenience and flexibility, is the extent of the available internal memory, or *RAM*[10] (for "Random Access Memory"). A minimum configuration today would include *at least* 5 MBytes of RAM, preferably even more (i.e., ca. forty times the amount of memory available in the original MACINTOSH!).

A few additional preliminary comments are in order at this point regarding the legendary "user-friendliness" of MACINTOSH computers (discussed in greater detail in Sec. 4.4). Some diehard critics continue to dismiss the idea of the *graphic user interface* ("GUI") as nothing more than a "cute gimmick", an unnecessary and even offensive intrusion in the life of the virtuous "power user". The same traditionalists characterize a mouse as an inefficient and awkward substitute for the tried-and-true (and perfectly adequate!) keyboard. Unfortunately, attitudes like these serve mainly as evidence of narrow-mindedness, shedding little light on the real issues. The genius of the innovative MACINTOSH approach to computing was that it enabled ordinary people to put computers to work for them almost immediately[11]—very effectively, and on a wide variety of tasks—exploding the myth that it takes a "special kind of mind" to enjoy the benefits of a word processor. Macintosh software available today is at least as versatile and powerful as software for

---

[10] English computer terminology, including "RAM", has been adopted virtually unchanged almost everywhere in the world. An exception is France, which has a long-standing reputation for resisting outside linguistic influences. The French refer to what we call RAM as «mémoire vive» ("living memory"), in contrast to the other type of computer memory, «mémoire morte» ("dead memory")—"ROM" in English. *ROM* (for "Read-Only Memory") is permanent memory from which information can only be withdrawn. The content of a computer's ROM is determined by the manufacturer, and only by installing new "chips" can ROM-resident data be augmented or modified.

[11] A special magazine supplement to the *Handelsblatt* (*Handelsblatt Magazin 2/86*) included the observation: "Even absolute computer novices find themselves able in a few hours to show tangible progress in data processing with the MACINTOSH."

the MS-DOS environment—and much less challenging to master. With respect to the alleged inefficiency and awkwardness of mouse-based operation it is important to recognize that there are very few situations in which use of a mouse is *mandatory*—most MACINTOSH (and WINDOWS) programs feature a comprehensive array of time-saving keyboard shortcuts. Nevertheless, a mouse can be a most welcome companion, especially in the early stages before one has developed a confident command (or interest in) the various keystroke alternatives, and it is invaluable later as a backup if one suddenly becomes unsure of the precise form of a particular keyboard command. The latter argument is a particularly compelling one for those users who work intermittently with a wide variety of programs—most desktop publishers, for example!

There is admittedly one (rarely discussed) detrimental side-effect of a "friendly" user interface: experienced users develop a tendency to explore the features of a new program exclusively by "doing" rather than by reading the manual. This is a perfectly natural reaction, since nearly all MACINTOSH programs are structured in rather similar ways; indeed, their use soon becomes quite intuitive. In one sense the phenomenon serves to confirm the greatest strength of a fully developed GUI system: an average (experienced) user can sit down at the keyboard and within minutes enjoy the first evidences of success with a new program. But it also constitutes a serious problem, because it encourages the MACINTOSH user to become lazy, and to forego the serious attention to program manuals and other documentation that is in fact the key to real proficiency and maximum utilization of a set of resources, leaving much of the true potential of MACINTOSH software sadly untapped.

We conclude this introduction to the MACINTOSH by noting briefly a few options for enhancing a basic set of hardware. One very important feature of all MACINTOSH computers is the ease with which relatively inexpensive third-party memory chips can be introduced. Indeed, the RAM complement can be expanded in some cases to as much as 68 MBytes. More memory translates into greater flexibility, including the possibility of loading several applications programs simultaneously. Most MACINTOSH models also provide "slots" capable of accepting various types of accessory boards ("cards"), including color-graphics boards, processor boards (the "Mac 286" card, for example, which can transform any MACINTOSH temporarily and on demand into the equivalent of an MS-DOS/AT system), accelerator boards (for increasing computation speed), network cards (including support for "Ethernet"), and A/D and D/A converters.[12]

Mention should also be made of flexible expansion devices (from Texas Instruments, for example) that provide access to as many as six additional "slots" for cards such as the ones already described.

4.3 IBM-Compatible Computers

The major alternative to a MACINTOSH computer as a basis for desktop publishing is a personal computer ("PC") descended from the one first introduced in 1981 by IBM. The IBM-PC represented one of the first serious attempts to promote computers as valuable tools for use in home and small-office environments, and the PC and its successors (the vast majority of them manufactured by companies other than IBM) have been enormously successful. Indeed, it is almost essential that a generic computer carry the label "IBM-compatible" if it expects to have a chance of surviving in the marketplace.

The feature that most clearly distinguishes what is commonly referred to as a PC from a MACINTOSH is the software at its heart: the computer's operating system. All IBM-compatible computers operate under a common system that began life as an in-house development project known as "QDOS" at Seattle Computer Products in 1980. This system was later purchased by Microsoft, which had accepted the challenge of providing IBM with an operating system. QDOS was soon transformed into "PC-DOS version 1.0" for installation in the first PCs.[13] Essentially the same operating system was subsequently made available by Microsoft to other computer manufacturers under the name "MS-DOS" (i.e., "Microsoft DOS"), although each DOS

---

[12] "A", analog; "D", digital.
[13] "DOS" is the acronym for "Disk Operating System", signifying that essential features of the computer's user interface are derived from disk-based files rather than ROM chips permanently installed in the computer.

implementation required minor adaptation to reflect hardware details unique to the design of a particular computer.

MS-DOS has of course evolved considerably since its first release in 1981, but it has changed relatively little in its overall structure. Thus, version 1.1 (1982) added the capability of addressing double-sided diskettes, and version 2.0 (released the following year) was designed to be applicable to systems with one or more hard disks, including provisions for a hierarchical file system (cf. Sec. 5.2). This gradual evolution has continued, with the most recent release of MS-DOS being version 6.2.

MS-DOS can best be described as a *command-driven* operating system. Thus, when the computer is first turned on and the operating system has been loaded, what appears on the screen is nothing more than a *system prompt* of the form "A>" (where "A" refers to a specific disk drive). At this point the user is expected to issue from the keyboard some specific instruction, expressed in a highly structured way that the operating system is capable of understanding. For example, issuing the command

TYPE C: \biblio\blie.txt

would cause the computer to attempt to display on the screen the contents of a particular file with the name "blie.txt"[14] located in a subdirectory called "biblio" on a disk drive that has previously been designated as "drive c". As soon as 23 lines of the corresponding text have appeared (the maximum number of lines that can be accommodated on an MS-DOS screen) the first lines will "scroll" rapidly off the top to make room for more text at the bottom. Suspending text output long enough to read the emerging message requires the user quickly to press the key combination "Ctrl–NumLock"; output is subsequently restored by pressing any key on the keyboard.

Commands of this general type are used to initiate every desired action, such as producing a list of all the files on disk A (command: "DIR a:"), erasing the directory record(s) of one or more files ("DEL c:\biblio\blie.*"; this would in fact eliminate from the location specified *all* files characterized by the name "blie" irrespective of their filename extensions), or copying a disk ("DISKCOPY"). DOS version 3.3 includes about 70 such com-

---

[14] An MS-DOS file name is limited to a maximum of eight characters, which may be followed by a period and up to three additional characters constituting a "filename extension", often used for purposes of classification. By contrast, MACINTOSH file names may consist of as many as 31 characters.

mands, each of which must be expressed in precisely the correct way—often with an appropriate set of parameters and modifiers—if the desired result is to be achieved. Fortunately, the average user rarely has occasion to issue more than perhaps a dozen of the most important commands, resorting to the DOS manual or some less intimidating reference source (e.g., Simrin, 1988) whenever a less familiar operation must be performed.

An *executable file* (i.e., an applications program) is initiated by typing its name at the system prompt. Unlike the MACINTOSH operating system, DOS does not provide a standard interface for utilization by all applications programs; each program is instead responsible for establishing its own interface. As a result, it is rare to find two programs that behave in the same way (although certain programming conventions have become relatively common). Similarly, input from a mouse (cf. Sec. 4.4) is not a feature supported directly by DOS, requiring installation of separate "mouse software" such as MICROSOFT MOUSE or EXPERT MOUSE (from Kensington).

Several attempts have been made to make DOS-based computers behave in a more "user-friendly" way. The most familiar is the MICROSOFT WINDOWS graphical "shell" utility that imposes a MACINTOSH-like operating environment between the user and the DOS internal structure.[15] Once WINDOWS is installed, users with access to WINDOWS-based software acquire many of the advantages of a graphics-based user interface, including active pictographs representing programs and files, pull-down menus, and full mouse support, as well as ready access to high-quality printer fonts and reasonably accurate display of the way a document will appear in print. This convenience has its price however: WINDOWS consumes an impressive amount of hard-disk space, and the fact that it is essentially an artificial "fix" applied to a non-graphical operating system results in very complicated procedures for installing (and removing!) the various WINDOWS applications.[16] Indeed, it was only with WINDOWS version 3.1 (1992) that most critics were con-

---

[15] WINDOWS is in fact so "Macintosh-like" that it was the subject of a prolonged and bitter legal battle between Apple and Microsoft, with Apple claiming (unsuccessfully) that WINDOWS violated copyright restrictions.

[16] "WINDOWS provides those training wheels that are the hallmark of a graphical environment, but you have to be a unicyclist to get rolling in the first place"; J. Heid, "Mac vs. PC: A Critical Comparison of the Macintosh and IBM PC Worlds", *Macworld*, March 1991, pp. 120–129.

vinced that a reasonable approximation to the convenience of the MACIN-
TOSH working environment had at last become available to PC users.[17]

Major improvements were promised for a redesigned version of WINDOWS,
known as WINDOWS NT (for "New Technology"), but the latter has not yet
lived up to expectations, in part because of a lack of applications software
written to take advantage of its unique characteristics.[18] WINDOWS NT is in
fact a completely independent operating system, no longer subordinate to
MS-DOS, featuring full *multitasking* (simultaneous processing of multiple
tasks) as well as a revolutionary new approach to graphic output that the
developers hoped might make PostScript obsolete. Unfortunately, WINDOWS
NT requires the availability of at least 30 MBytes of free hard-disk space,
with a recommended reserve of 100 MBytes;[19] it apparently also has the
disadvantage of greatly reducing the speed at which ordinary WINDOWS
software operates.[20]

Three major reasons have often been given for preferring MS-DOS systems over their MACINTOSH counterparts: lower initial cost, greater processing speed, and wider corporate acceptance. As noted previously, the first no
longer provides a compelling argument, especially since a MACINTOSH system
normally includes a number of useful features that must be added separately
(at considerable expense) to a PC. Speed distinctions are also much less
significant than in the past, particularly if one's DTP activities are not heavily
dependent on working with complex color graphics. The newly released
POWER MACINTOSH computers ("POWERPCs") contain RISC ("Reduced Instruction Set Computing") processors that are actually faster (and much
smaller and less expensive) than the highly publicized, "super-fast" Pentium
processor developed by Intel for IBM-compatibles.[21]

---

[17] In a recent article providing advice on the purchase of an IBM-compatible computer for desktop publishing purposes (in a relatively unbiased journal) the author
observes that "…WINDOWS is now almost half as good as the MACINTOSH system
software"; O. M. Kvern, "More Power to You: Buying an IBM-Compatible Computer in an Age of Uncertainty", *Aldus Magazine* **1993**, *4* (7), 44–48.
[18] A similar problem has plagued IBM's own alternative to DOS, OS/2.
[19] J. Borzo and R. Luhn, "Are They Just NT Promises?", *Publish*, October 1992, pp.
70–76.
[20] S. Lininger, "The Perfect Publishing PC", *Publish*, April 1994, pp. 64–66.
[21] MACINTOSH "POWER PCs" are the subject of several articles in the May 1994
issues of *Macworld* and *MacUser*.

Nevertheless, many people still choose to follow the PC route to desktop publishing, often because of past experience or close contact with others operating on the basis of this platform. A minimum hardware configuration in this case should probably include an Intel 80486 (or Pentium) processor operating at no less than 50 MHz, a 200-MByte hard disk (preferably with a SCSI interface for increased speed and flexibility), at least 8 (some would argue 64) MBytes of RAM (in the form of easily supplemented SIMMs— "Single Inline Memory Modules", which are a standard feature of MACINTOSH computers), and provisions for both 3.5″ and 5.25″ floppy disks.[22]

4.4 *The* MACINTOSH *Operating System*

Having discussed briefly in the preceding section the MS-DOS operating system it is appropriate that we turn now to a brief introduction to the MACINTOSH operating system, many features of which are also a part of the WINDOWS operating environment for the PC. As previously suggested, computer software can be divided into two fundamental categories:

- *System software* (or the *operating system*), consisting of one or more programs that supervise and monitor basic operations of the computer itself (including data transfer, file management, and implementation of other programs)

- *Applications programs* to assist the user in manipulating information: programs for the creation and modification of files containing text and graphics, acquisition of experimental data, mathematical computation, and numerous other tasks

A significant portion of the operating system may be permanently resident in the computer's ROM, but a MACINTOSH computer becomes fully operational only after an organizational program called SYSTEM has been loaded into RAM from a diskette or a hard drive. Most MACINTOSH applications

---

[22] These suggestions have been adapted from a list provided in the article by O. M. Kvern cited in note 17 on p. 124. A somewhat more technical source of advice is the article by S. Lininger cited in footnote 20.

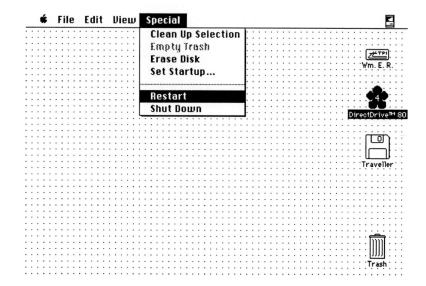

Fig. 4–1. The MACINTOSH "desktop" screen display, showing one of several "pulldown menus".

also require the availability of a second system program called FINDER,[23] which facilitates access to stored data while ensuring efficient management of all available data files and applications programs.

Once a set of system software has been successfully loaded, the screen display provides a signal to the user that the computer is ready to operate. In the case of the MACINTOSH what appears is a set of images known collectively as the "desktop" (Fig. 4–1). In effect this is a visual communications interface consisting of a panel across the top of the screen known as a "menu bar", a series of pictograms ("icons") representing the available disk drives, and a "wastebasket" for disposing of unwanted files. The menu bar provides access to various control functions and information sources that the system software places at the user's disposal, acting as a convenient substitute for memorized commands entered from the keyboard.

---

[23] A more complex version of the finder, called MULTIFINDER, makes it possible to load and access several programs simultaneously—a tentative step in the direction of multitasking. The most recent major revision of the MACINTOSH system software, "System 7", operates exclusively in "multifinder mode".

At this stage of operation commands are normally issued to the computer with a device known as a *mouse*. This is a special input tool whose principal component is a freely rotating ball projecting a few millimeters below the under side of a small box, the top of which reveals one or more large pushbuttons.[24] Moving the mouse across any flat surface causes the ball to rotate, and this in turn induces a *mouse cursor* to trace an equivalent path across the display screen. The cursor can take various forms under various circumstances, but it initially looks like an arrow pointing up and to the left. Placing the cursor over one of the items in the menu bar and pressing the mouse button produces a "pulldown menu" (cf. Fig. 4–1), from which any listed item can be selected (activated) by pointing at it with the cursor and then *releasing* the mouse button. Under other circumstances the mouse cursor can be used for marking or otherwise operating upon a specific piece of text (one letter, a few words, or an entire document), selecting and moving individual components within a complex illustration, and much more.

Pressing the mouse button twice in rapid succession ("double-clicking") while the cursor is located over one of the *drive symbols* brings to the screen a visual record (directory) of the contents of that drive, with the information displayed either as a set of icons (for applications programs, data files— even "file folders", which are the MACINTOSH equivalent of subdirectories; cf. Fig. 4–2) or as a more informative list, which can be arranged in any of several ways (e.g., Fig. 4–3).

An applications programs—a word processor, perhaps—is activated ("loaded") simply by double-clicking on the corresponding screen icon, or on the icon of a document previously prepared with that program. The latter technique for loading a program provides immediate access to the selected document.

Most MACINTOSH program operations are initiated with menu commands that in turn apply to the contents of various types of "windows" (segments of the display screen reserved for data presentation, communication with the operating system, short-term data storage, etc.; see below). Menus are generally of the "pull-down" or "window-shade" variety, and appear only upon request, an expedient that conserves valuable display space (cf. Fig. 4–1).

As noted above, a *window* is a discrete portion of the screen display, usually rectangular, that serves a specific purpose, such as facilitating interac-

---

[24] A MACINTOSH mouse has only a single button, whereas an MS-DOS mouse is equipped with two or even three pushbuttons.

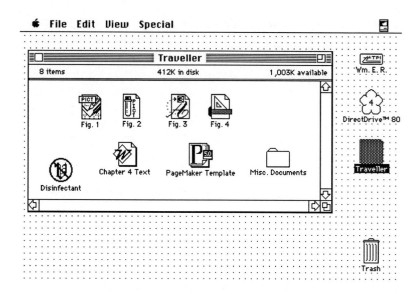

Fig. 4–2. Content display for a MACINTOSH floppy disk, presented in "icon" format.

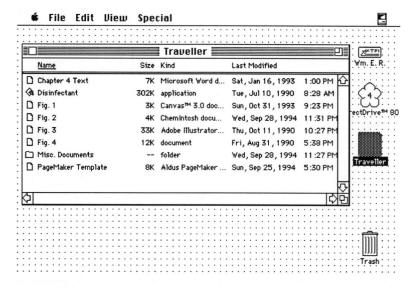

Fig. 4–3. The disk directory from Fig. 4–2 presented as an alphabetically arranged list. Other options include display by date, size, or type of file.

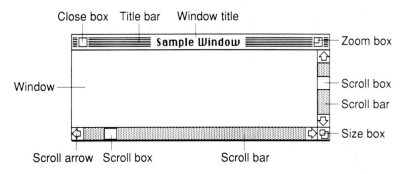

Fig. 4–4. The structure of a typical MACINTOSH "window", with both horizontal and vertical "scroll bars".

tion with the operating system (cf. Fig. 4–6, which shows a window for activating a particular printer) or permitting the introduction of structured text (cf. Fig. 9–1). Most windows are receptive to some type of input from the mouse or the keyboard (cf. Figures 4–5 through 4–9). A typical MACINTOSH window is surrounded by a *frame* (Fig. 4–4) that also incorporates a number of control functions. For example, if one directs the mouse cursor to the upper portion of the frame (the *title bar*) and then "drags" the mouse while its button is depressed, that window will "move" in a corresponding way across the screen. Horizontal and vertical *scroll bars* along the bottom and right-hand edges of the frame allow one to move the "scene behind the window"—revealing, in the case of a document window, portions of the corresponding document that would otherwise have been "cut off" by the frame.

One unique feature of the MACINTOSH operating system is the "apple menu", a pull-down menu always accessible at the left-hand edge of the menu bar and symbolized by a small representation of an apple (the manufacturer's logo; cf. Fig. 4–1). This menu provides direct, immediate access to a set of special (relatively small) applications programs and functions known as *desk accessories (DAs)*. The nature of the particular set of available DAs is determined by the user, in some cases with the aid of a utility program (e.g., Apple's FONT/DA MOVER, Fig. 2–6, or SUITCASE from Symantec).[25] Calling

---

[25] Version 7 of the MACINTOSH system software provides a more straightforward approach to installing DAs (and fonts), eliminating the need for the FONT/DA MOVER.

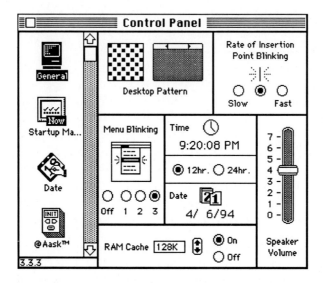

Fig. 4–5. The MACINTOSH CONTROL PANEL desk accessory, used for establishing various parameters associated with the operating system and certain auxiliary programs.

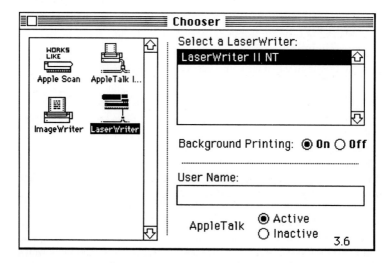

Fig. 4–6. Display associated with the CHOOSER desk accessory for designating an "active" output device.

up a desk accessory from the apple menu temporarily interrupts communication with other programs that are currently active, but the state of such programs is in no way permanently altered.

Several DAs are included as part of the MACINTOSH system software, and many others are available commercially. A remarkably wide variety of useful DAs has also been made available within the public domain via cooperating *user's groups*,[26] on-line data networks, or the authors directly. Public-domain software is of two types: programs created by public-spirited individuals who provide them to others at no cost *(freeware)*, and programs offered for a modest, often voluntary, fee payable directly to the developer on an "honor system" basis *(shareware)*.

There are now several hundred Macintosh desk accessories, some of which are extremely valuable. Examples of "official" Apple system DAs include:

- the CONTROL PANEL (Fig. 4–5), which is used for adjusting certain operating parameters of the system, including the computer's understanding of the current date and time and the responsiveness of the mouse and keyboard

- CHOOSER (Fig. 4–6), used for designating the particular printer that is to be regarded as active

- FINDFILE (Fig. 4–7), a relatively primitive tool for locating a particular file from among the many that may be dispersed throughout the complicated multilevel file system of a large hard disk

- KEYCAPS (Fig. 4–8), which provides a convenient graphic representation of the keyboard as a way of facilitating the identification of key combinations leading to various special symbols available within a particular font

Many other useful DAs are available from a variety of sources. Examples include:

- DISKTOP (cf. Fig 5–5), one of several programs enabling the user at any time to ascertain the status of a particular file, and to discard files or alter the characteristics of files not currently active

---

[26] An example of a large and very active MACINTOSH user's group is "BMUG" (the Berkeley Macintosh User's Group, with headquarters in Berkeley, California), which publishes a comprehensive catalog of available shareware (Potkin, Hansen, and Schneider, 1992).

Find File

⊂▭ DirectDrive™ 80

Search for: logo

1.3

▢ LOGO
▢ VCH Logo (4/90)
▢ Ullmann Rev. Logo
▢ **Ullmann's Logo**
▢ Ullmann Rev. Logo PICT

Created:	Tue, Jan 21, 1992; 8:59 PM
Modified:	Sun, Jun 27, 1993; 8:33 PM
Size:	4798 bytes; 6K on disk

⊟ **Ullmann Folder**
⊟ VCH
⊂▭ DirectDrive™ 80

Fig. 4–7. Apple's FINDFILE desk accessory for locating a particular file on a MACINTOSH disk.

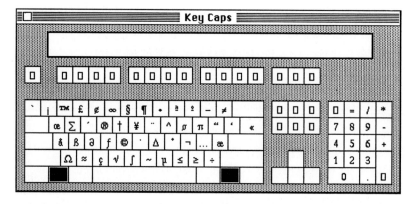

Fig. 4–8. KEYCAPS, a convenient MACINTOSH desk accessory for determining the keyboard locations of all the characters available in a particular font. The symbols shown are the ones associated with the "option" keys (one of which has been depressed, causing both of them to be highlighted in black).

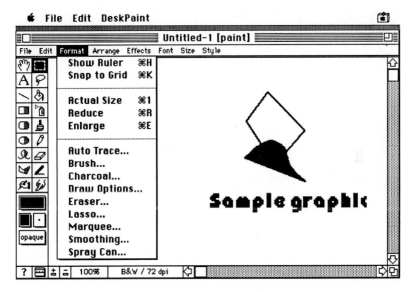

Fig. 4–9. User interface for the DESKPAINT desk accessory (Zedcor), showing the content of one of the program's pull-down menus.

- A full-function SCIENTIFIC CALCULATOR responsive both to the mouse and the keyboard

- DAs that offer remarkably extensive graphics capabilities, including DESKPAINT (Fig. 4–9) and DESKDRAW (both available from Zedcor)

- Specialized DAs such as MACEQUATION and EXPRESSIONIST for "typesetting" complex mathematical and physical equations (cf. Sec. 9.7)

One other important feature of the MACINTOSH operating system that should not be overlooked is the *clipboard.* The clipboard can be regarded as a special portion of the computer memory, virtually unlimited in extent, that can be called upon for the temporary storage of information one wishes to pass from one program to another (or transfer to a new location within the same program). Full compatibility with the clipboard is a feature of essentially all MACINTOSH software—complete programs as well as DAs—which means that it is an almost trivially simple matter to transfer data—text, numerical information, even graphics—back and forth between applications and files of every conceivable type.

4.5 *Networks and Communication*

Networking is a term used to describe the interconnection of several computers. Networks often have as their primary purpose providing several
workstations with access to a single "master" system, an arrangement similar
to that which has long been commonplace in the context of large mainframe
computers. A network makes it very easy for several users to share and exchange data, but it also leads to more efficient utilization of expensive peripheral devices, including laser printers and mass-storage facilities. In addition, networks often represent the best way of providing groups of people
with access to broader *public networks* (e.g., Internet; cf. Lane and
Summerhill, 1992, and Dern, 1994) as well as commercial on-line data services like COMPUSERVE or DIALOG (cf. LeVitus and Ihnatko, 1992, and especially Sec. 12.2.4), which in turn opens the way to enormous (external)
databases.

Another important advantage of networking is the potential it creates for
shared utilization of in-house databases. For example, preparing a new
product catalog, or even a company newsletter, may require that several individuals at different locations have access to various types of comprehensive and current information. Indeed, desktop publishing systems in corporate settings often reveal their full potential only after they have been
upgraded to include convenient and reliable communication with other in-
house data-processing facilities.

A *local area network* (*LAN*) is established by connecting several computers and their associated peripheral devices through some type of cable
system, and then installing appropriate communications software. *Closed*
LANs (such as APPLETALK) are usually restricted to equipment of a single
type (e.g. MACINTOSH workstations), whereas *open* LANs are adaptable to a
wide range of hardware, which may also be dispersed over a rather wide
area (e.g., the TOPS system, developed by Sun Microsystems). Networks can
also be *configured* in various ways, denoted by fanciful designations such
as "tree structure" and "token ring" (for more information see, for example,
Nance, 1993, Stallings, 1993, McNamara and Romkey, 1993, and Motorola
Codex, 1993).

Most devices associated with the MACINTOSH family (computers, hard
disks, printers, etc.) lend themselves to straightforward interconnection

directly via APPLETALK modular cables. MACINTOSH systems can be linked with equipment from the MS-DOS world by outfitting them with appropriate circuit boards (of the ETHERNET variety, for example), which in turn attend to the inevitable "translation" problems. An ETHERNET network permits each individual user to act both as a "server", making data available to others, and as a "client" able to receive information from other participants in the network. Individual users are usually free to designate portions of their personal databases as "closed" to access from outside. A more flexible technique for controlling data access is establishment of a system of *passwords*: "secret" code words that must be correctly entered into the system before one is permitted to examine or alter particular files.

Networks represent only one of the possibilities for establishing communication between MACINTOSH equipment and MS-DOS computers. All MACINTOSH computers now come equipped with 3.5″ disk drives that are fully capable of reading data stored on 3.5″ MS-DOS disks, and special (external) drives are available for working with 5.25″ MS-DOS disks. Such "foreign" data files are "imported" and thereby adapted to suit the MACINTOSH environment with the aid of a system program called APPLE FILE EXCHANGE. In fact, one can even *run* MS-DOS programs on an ordinary MACINTOSH if the need ever arises. One way of accomplishing this feat is by first activating a program from the SOFTPC family (from Insignia Solutions), leading to a special MS-DOS window on the MACINTOSH screen that looks and behaves exactly like an ordinary MS-DOS (or WINDOWS) display. No special hardware whatsoever is required, and SOFTPC works remarkably well under a wide variety of circumstances. With MULTIFINDER active it is even possible to run one or more MACINTOSH programs simultaneously with SOFTPC, and the MACINTOSH clipboard permits facile data transfer between these two otherwise incompatible computer worlds. The major disadvantage of SOFTPC is that some MS-DOS programs run rather slowly in the artificial environment due to the multitude of complex "mental gyrations" the computer is required to perform.

4.6 *Computer Viruses*

It would be a mistake to conclude this introduction to computers without brief mention of one irksome problem that has surfaced within the last few years: *computer viruses* and their equally unpleasant cousins, *worms* and *Trojan horses*.[27]

Unlike its biological namesake, a computer virus is a human invention, deliberately conceived for destructive or disruptive purposes. In essence, a computer virus is a piece of software that replicates and then attaches itself to other programs and data files. Moving an infected file from one computer to another has the effect of implanting the virus in the new environment as well, where it can again multiply and develop into a potential source of further infection.

Viruses made their first appearance more than a decade ago, the product of one or more warped but clever minds, and they spread like wildfire, thanks in large measure to the tradition among computer users of generously sharing their personal software resources. Even commercial computer networks fell victim to the plague, and many users have suffered infections as a result of files innocently downloaded via personal modems.

Fortunately, most viruses can be characterized as nuisances rather than threats—causing "humorous" messages to appear unexpectedly on the screen, for example, or erratic "beeping" noises. Others are far more serious, leading to the mysterious disappearance of files, or even to the erasure of an entire hard disk. Even the most benign virus represents a potentially serious problem, however, since it may well interfere with crucial parts of a computer's operating system, producing unstable behavior and inexplicable "system crashes".

The best strategy to adopt with respect to computer viruses is analogous to that directed against biological viruses: a combination of prophylaxis and medication. Above all, be wary about introducing anyone else's disks into your computer! No virus can enter a computer by itself: infection always

---

[27] A "worm" is a program that replicates and spreads in much the same way as a virus, but it does not become attached directly to other software. Most worms are spread through computer networks. A "Trojan horse" is a program (non-replicating) that pretends to do one thing while actually performing in secret some very different and dastardly deed.

starts with a previously infected file. The risk of acquiring an infection from a commercial program or data file is negligible, since distributors take extreme precautions to protect their products. The real danger lies almost entirely in the shared disk—or a personal disk innocently inserted into the (infected) computer in one's office. (Merely *inserting* an "unlocked" disk in an infected computer can result in a corrupted directory file!)

Adopting a "conservative lifestyle" with respect to one's own computer and floppies is good prophylaxis, but this should be combined with more active measures, like outfitting one's computer with powerful "antivirus" software. Many such programs are available for permanent installation on the computer's startup disk. Their role is to examine the existing system for evidence of infection, remove any problems that are discovered, and screen all disks inserted in the future, issuing an unmistakable warning at the first sign of trouble. Highly regarded antiviral programs for the MACINTOSH include DISINFECTANT (available without cost from John Norstad at Northwestern University), SAM (Symantec) and VIREX (Datawatch). Analogous products for the PC include NORTON ANTIVIRUS (Symantec) and ALLSAFE (XTree).

5 Data Storage

5.1 General Considerations

Adequate data-storage capacity is an absolute prerequisite if one proposes to work extensively with electronic text and illustrations. The act of turning off a computer effectively clears everything out of its memory, so one obviously needs some alternative way of preserving recent accomplishments for future use. The least expensive and most portable storage medium at the present time is the *diskette* (*floppy disk*; cf. Sec. 5.2). Apart from limited capacity, the chief disadvantage of relying exclusively on diskettes is their relatively long *median access time* (cf. Sec. 5.3). Data transfer can be achieved much more rapidly to and from a *hard disk* (Sec. 5.3). Hard disks also provide vastly more storage space than diskettes, but they are relatively expensive. Other alternatives include *tape drives* (which, though not suited to efficient random data access, do offer excellent emergency backup for programs and data files) and a more recent innovation, the *optical storage device*,[1] which will be discussed only rather briefly (in Sec. 5.4) since it has just begun to play a significant role for the individual desktop publisher.

---

[1] The latter category includes CD-ROM, WORM, magneto-optical, and floptical drives. All three share the characteristic of utilizing a laser beam at some stage during the reading and/or writing of data. The cost/benefit balance with respect to data storage is likely in the near future to shift dramatically in the direction of optical devices, although at this point some aspects of the technology still suffer from a lack of standardization.

5.2 Diskettes

Diskettes or *floppy disks* are flexible plastic disks coated on both sides with a thin layer of an iron-containing substance whose particles are subject to selective magnetic alignment. During active use the disk rotates continuously at a speed of 300–600 revolutions per second. Most personal computers utilize diskettes that are either 5.25″ or 3.5″ in diameter, with the latter gaining rapidly in popularity.

Data storage with a diskette is accomplished by causing a *read/write head* located within the disk-drive mechanism to generate a data-specific fluctuating magnetic field that in turn causes iron particles on the disk surface to become magnetized in a particular way. In the reverse process, weak magnetic fields surrounding the particles evoke electrical signals in the head, permitting stored information to be "read" from the diskette. In the case of a 3.5″ drive the read/write head gains access to the stored data through a small *window* (cf. Fig. 5–1) in the plastic diskette case, which is protected by a spring-loaded sliding shutter. The 5.25″ diskettes most often associated with IBM-compatibles tend to be somewhat more vulnerable, since they are stored in less robust containers with permanent cut-out windows for data access.

Many manufacturers produce diskettes, and to a wide variety of specifications subject to more or less strict quality-control standards. An important distinction is made between *single-sided* and *double-sided* diskettes. In re-

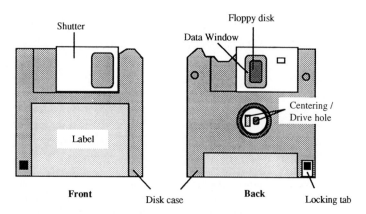

Fig. 5–1. The physical structure of a 3.5″ diskette.

ality, all diskettes are coated on both sides with magnetic material, but single-sided diskettes are only *certified* with respect to one side, so the second side may well exhibit defects. A further distinction divides diskettes into the categories *single-density* (SD), *double-density* (DD), and *high-density* (HD), characterized by increasing numbers of magnetic particles per unit of surface area and a correspondingly greater potential for capturing information.

A blank diskette as it comes from the manufacturer represents only a *potential* carrier of information. Before it can be utilized it must first be *formatted (or initialized)*, in the course of which an essentially random magnetic surface is organized into a series of concentric circles *(tracks)*, to each of which an appropriate computer system will subsequently enjoy nearly instantaneous access. A particular track constitutes a narrow circular band of magnetic material that is assigned a unique identification number *(track number)* within the range 0 to N (where $N = 79$ for a standard MACINTOSH diskette). Track 0 is located near the outer edge of the diskette, and track N is the one nearest the center. A shift of activity from one track to another requires that the drive mechanism effect a precise lateral displacement of its read/write head. The *density* of the data tracks is usually expressed in units of tracks per inch (tpi). For example, a typical diskette drive utilizes a density of 135 tpi, whereas most removable hard disks (cf. Sec. 5.3) are engineered for a track density of ca. 350 tpi.

Tracks are in turn subdivided into zones known as *sectors* (cf. Fig. 5–2). A small portion of each sector is reserved for identification purposes, with the remainder available for data storage. While the distance between a given track and the center of the disk is a fixed quantity, invariant from one disk to another, the location of a particular sector is a function of the relative position of a *centering hole* cut into the diskette itself.

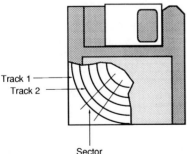

Track 1
Track 2

Sector

Fig. 5–2. Data structure of a 3.5″ diskette.

Some diskette drives (e.g., those employed with IBM-compatible computers) operate with a constant rotational speed, which means that the density of information storage must increase from the periphery of a diskette toward its center. The same is true for "high-density" (1.44 MByte) diskettes in the MACINTOSH environment,[2] but the traditional 400- and 800-KByte MACINTOSH diskettes are variable-speed storage media, with the proper rate of rotation a function of the identification number of the currently active track. As a result, more information can be stored in the longer tracks near the periphery, and roughly constant information density is maintained across the entire diskette surface. The number of sectors within a particular track on a standard MACINTOSH diskette is thus a function of the distance of that track from the center. The overall storage capacity for such a diskette exceeds that of a diskette operating at constant rotational speed.[3] As indicated previously, information storage need not be limited to one side of a diskette provided the drive unit is equipped with two read/write heads. This permits a doubling of the effective storage capacity (from 400 KByte to 800 KByte in the case of a standard MACINTOSH diskette). The *mean access time* (cf. Sec. 5.3) for retrieving data from a diskette is approximately 100 ms.

In order to locate a particular piece of data on a diskette (or a hard disk; cf. Sec. 5.3) the computer's operating system must make reference to the disk's *directory*. This "table of contents" is stored directly on the disk itself, and contains a record of the name, type, size, creation date, and location of each stored document (cf. Fig. 5–3). *Deletion* of a data file in most cases amounts to nothing more than eliminating the associated directory entry; the *data* constituting the file would normally disappear only later, when the corresponding disk sectors were called upon to house data from a new file. Nevertheless, the act of clearing a directory is virtually equivalent from a practical point of view to reestablishing a blank (albeit formatted) disk, be-

---

[2]   Apple began offering MACINTOSH users the convenience of the more efficient HD diskettes with the release of the model SE-30 computer. Since that time all MACINTOSH computers have been equipped with the so-called SUPERDRIVE, capable of distinguishing between and reading (or writing) 400-KByte, 800-KByte, and 1.44-MByte diskettes, including 3.5″ 1.44-MByte (and 720-KByte) diskettes formatted under MS-DOS.

[3]   The 80 tracks on a single-sided 400-KByte MACINTOSH diskette are divided into sectors as follows: 12 sectors per track for the first 16 tracks, 11 sectors in each of the next 16 tracks, then 10, 9, and finally 8 sectors per track. The 80 tracks thus provide a total of 800 sectors, each with a storage capacity of 500 Bytes.

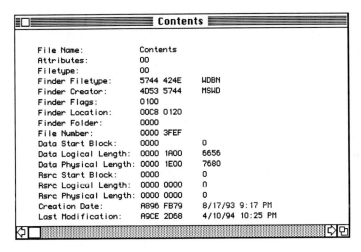

```
▤□▬▬▬▬▬▬▬▬▬ Contents ▬▬▬▬▬▬▬▬▬

   File Name:            Contents
   Attributes:           00
   Filetype:             00
   Finder Filetype:      5744 424E   WDBN
   Finder Creator:       4D53 5744   MSWD
   Finder Flags:         0100
   Finder Location:      00C8 0120
   Finder Folder:        0000
   File Number:          0000 3FEF
   Data Start Block:     0000        0
   Data Logical Length:  0000 1A00   6656
   Data Physical Length: 0000 1E00   7680
   Rsrc Start Block:     0000        0
   Rsrc Logical Length:  0000 0000   0
   Rsrc Physical Length: 0000 0000   0
   Creation Date:        A896 FB79   8/17/93 9:17 PM
   Last Modification:    A9CE 2D68   4/10/94 10:25 PM
```

Fig. 5–3. Screen display of the directory information associated with a MACINTOSH disk file (displayed using MACTOOLS, Version 6.3, from Central Point Software).

cause the computer no longer has a record of the way in which the residual data are structured.[4]

The MACINTOSH operating system recognizes two alternative organizational schemes for a set of disk files:

● the original MACINTOSH file system, known as "MFS", and
● the more advanced hierarchical file system, "HFS", introduced in 1986.

---

[4] Strictly speaking, such a disk is of course *not* blank. Even when directory entries have been removed—or an entire directory is destroyed—the potential exists for recovering much of the apparently "lost" information. Thus, special programs have been developed for analyzing and applying a kind of emergency "first aid" to a disk that has been damaged, or that might still contain "deleted" files. An example of a program of this type for the MACINTOSH, distributed by Apple, is actually called FIRSTAID. More powerful "disk editors" include that supplied with NORTON UTILITIES (Symantec). A disk editor typically begins by treating the target disk as if it were a jigsaw puzzle, first identifying individual (labeled) pieces, and then characterizing, rearranging, and assembling these pieces insofar as possible into a consistent whole. Just as a missing table of contents for a book might be reconstructed by searching through the pages for chapter titles and section headings (or fragments, in the case of a damaged book), so a disk editor examines all the sectors on a diskette, one at a time, and creates a record of what it finds, on the basis of which it attempts to create a new directory.

A disk structured according to the *MFS system* recognizes no hierarchy whatsoever with respect to stored documents; all files reside at the same "level" within the disk directory. The *user* may choose to associate individual documents with specific "file folders"[5] in an attempt to establish some degree of order, but these folders play no role in the *computer's* organizational scheme as reflected in the corresponding directory. For this reason it is not possible to assign identical names to two different documents on a given MFS disk even if the documents themselves have been placed in different folders; any attempt to do so simply leads to an error message. This is not true with disks organized according to the *HFS system*, which was developed in response to the new demands imposed by hard disks and other storage devices capable of holding massive amounts of data. HFS folders correspond to true, independent subdirectories.

Under both systems it is a straightforward matter to reorganize folder content and create new folders—even several levels of "folders within folders"—providing an incentive for the user to optimize the data-storage structure for each disk. For example, one folder might be reserved exclusively for documents created by the word processor MICROSOFT WORD (with the "type" designation WDBN[6]), while another could be used for documents of the type DRWG created under MACDRAW. Alternatively, files of various types but all related to a particular project might be grouped together in a single folder. With an HFS disk arranged according to such a *tree structure* there is no restriction against several documents sharing the same name so long as they are kept in separate folders (cf. the documents called "E2" in the folders labeled "Documents 4/91" and "Drafts" in Fig. 5–4). In theory, a MACINTOSH disk structured according to the HFS system might contain as many as 15 000 separate files, whereas an MFS disk is limited to ca. 800 files.

---

[5] "Folders" are user-defined collections of programs, documents, and data files intended to establish order in what might otherwise become a very complicated environment; for an example of the way in which folders are depicted on the screen see Fig. 5–4. The "folders" on a MACINTOSH are equivalent to subdirectories under MS-DOS or "groups" and "items" in WINDOWS.

[6] The "document type" serves the MACINTOSH operating system as a guide to the relationship between a data file and the program with which it was created. *File extensions* perform a similar (though much more limited) service under MS-DOS or WINDOWS.

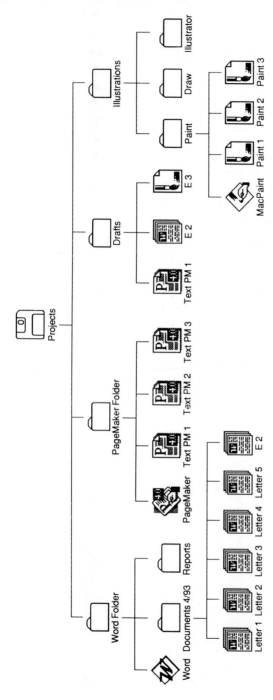

Fig. 5-4. An example of the way files might be organized under a hierarchical system.

We conclude this section with a few suggestions regarding the use and treatment of diskettes.

- It is always wise to purchase diskettes that are guaranteed to be of high quality; program and data files are extremely valuable, and it would be foolish to put them at risk for the sake of a few pennies saved.

- Backup copies—preferably more than one!—should be maintained for all important files (cf. Sec. 5.5). The security copies should be kept in a safe place far removed from the originals to provide added protection against loss due to fire, flood, or theft.

- Programs should always be loaded from backup copies, never from the original source diskettes. Replacing a damaged original can prove difficult and costly!

- Diskettes must be scrupulously isolated from magnetic fields of all types, including those associated with loudspeakers, electric motors, telephones, powerful halogen desk lamps, and transformers. Airport security systems have also been known to cause damage to computer disks.

- Diskettes of the 3.5″ variety are designed to be handled only by their plastic cases. One should never be tempted to open manually the sliding cover that protects the data window, and all contact with the magnetic surface must be strictly avoided. Even fingerprints can be a source of serious disk errors.

- It is good practice always to take advantage of a diskette's *write-protect* tab (or "lock"; cf. Fig. 5–1) as a way of preventing accidental erasure of important files (and also to minimize the chance of exposure to a "computer virus"; cf. Sec. 4.6).

- Diskettes should be stored in a dust-free environment protected from extreme temperatures and moisture, and preferably in an upright position. Never store disks in an automobile, in direct sunlight, or near a radiator.

- A disk label should always be filled out *before* it is affixed to a disk. If it becomes absolutely necessary to write on an already attached label one should use a felt-tip pen and apply minimal pressure in order to avoid damaging the fragile disk surface.

- Disk labels should never be erased because of the risk that residual eraser crumbs might damage the disk or even the drive mechanism.

● Dry labels are almost impossible to remove from a diskette, but a label carefully moistened with pure mineral spirits comes off quite easily, leaving almost no residue of paper or glue.

5.3 High-Capacity Magnetic-Disk Storage

Diskettes are suitable for storing only relatively small amounts of data. Large files and groups of files demand an alternative storage medium with greater capacity. The three mass-storage options currently applicable to personal computer systems are magnetic hard disks, magnetic tape, and so-called optical disks (Sec. 5.4).

Hard disk is a blanket term applied both to conventional (sealed) storage devices and the more versatile "removable" media. Hard-disk[7] data-storage systems function much like diskette drives, the principal difference being that the storage medium has a considerably greater capacity and is less accessible physically to the user. Hard disks obviously require larger directories than diskettes, but the data-management provisions are essentially equivalent.[8] A single disk-drive unit may actually incorporate several discrete disks, which are sealed within a dust-free environment and driven by a sophisticated precision drive mechanism. Some hard disks (*internal* hard disks) are permanently located in the case that houses the computer's central processor, whereas others (*external* hard disks) are accessories that can be added or removed at will. External hard-disk units vary in size from "stackables" that match the computer chassis to small, portable "pocket drives". Capacities range from as little as 40 MByte to 2500 MBytes (2.5 GBytes) or more.

The read/write heads of a hard disk unit are designed to swing back and forth along tightly restricted paths in very close proximity to the disks

---

[7]   The term "hard disk" is used indiscriminately to refer both to the storage medium itself and the drive *unit*, of which the actual disk constitutes only one part.
[8]   One minor difference is the fact that hard disks are often "partitioned" into multiple "volumes", each of which is treated like a separate storage medium.

themselves, although it is crucial that the head and the disk never make direct physical contact.[9]

Three factors (in addition to greater capacity) are primarily responsible for the popularity of hard-disk primary storage relative to diskettes:

● reduced *access time* (the delay associated with positioning the read/write head above a particular track; typical mean values range from 10–80 ms)

● brief *latency periods* (intervals during which a head is forced to await the arrival of the appropriate data block); these are a function of the rotational speed of the disk (typically 3600 min^{-1}), and usually average about 8 ms

● rapid *data transfer*—referring to the rate at which information passes back and forth between the computer and the disk. This is again somewhat dependent on rotational speed, but it is also a function of the *data density*, measured in bits per inch per track, as well as the nature of the computer *port* (the point of access on the computer for establishing a cable link to the hard-disk drive unit).[10] Values between 0.6 and 4 MBytes per second are common

The higher capacities and shorter access times for hard disks relative to diskettes are due mainly to:

● the potential for incorporating multiple disks into a single drive unit (typically three or more, together with a corresponding number of read/write heads)

---

[9] Hard drives long had a reputation for vulnerability due to the close tolerances involved in their construction. Even a minute dust particle falling between the read/write head and the rapidly rotating disk might disrupt the air cushion separating the head from the disk, leading to a "headcrash" and the loss of a considerable amount of data. Headcrashes were once rather common occurrences, but the risk of such an event has been greatly reduced by the introduction of more highly resistant disk surfaces and housings with tighter seals. Damage might also be caused by vibrations set up in the course of moving a hard-disk unit from one location to another, but this can be prevented by a mechanism for securing (or "parking") the heads whenever power is removed from the unit. Most disk drives now feature "self-parking" heads that lock automatically when the drive is shut down.

[10] Most MACINTOSH hard disks (and many IBM-compatible units) are linked to the computer through what is known as a "SCSI port" (an acronym for "Small Computer Standard Interface"), a gateway capable of exchanging information at a very rapid rate.

- high rotational speeds (which may be constant or variable)

- an increased number of tracks, which may have equal or varying numbers of sectors

- extraordinary precision in the construction of the drive mechanism

Removable hard-disk drives combine the portability and reliability of diskettes with the increased capacity and speed of a hard drive. Individual high-density magnetic disks, typically with access times below 30 ms and capacities of 44–104 MBytes, are installed permanently by the manufacturer in special protective casings that can be inserted into an appropriate drive mechanism in much the same way as an audio cassette is introduced into a tape player. Disks of this type are obviously valuable for backup purposes, but they are also convenient for archiving large amounts of data and for transporting voluminous data from one location to another, especially in conjunction with such storage-intensive activities as scanning and graphic design, or the shared use of databases.

5.4 *Other Alternatives*

Another well-established type of mass storage device is the *tape streamer*, with a typical capacity of 150 MBytes. Access times in this case are necessarily high, because tape is a *serial* storage medium: data must be written consecutively on a single long run of tape. Tape streamers are therefore restricted almost exclusively to data-backup tasks, and they have never been particularly popular for personal computers.

Optical storage devices constitute the most recent entry into the personal computer data-storage market. The most familiar are *compact disks* (CDs, like those used for recording music, with a diameter of 12 cm), featuring data capacities in excess of 500 MBytes. In effect a CD constitutes a new species of read-only memory (ROM), one that can provide rapid, reliable, and interchangeable access to data stores comparable to the content of a large dictionary or a modest encyclopedia. Thus, a single CD-ROM disk might contain a collection of literary works, geographical data equivalent to a set of detailed maps of the United States, a national telephone directory, or a large collection of fonts or "clip-art" images. CD-ROM drives are now offered by several manufacturers, and some computers already include them as standard equipment. The list of CD-ROM data collections available commercially (both for the MACINTOSH and IBM-compatibles) is growing rapidly, and now claims several pages in almost every software catalog.

Other forms of optical storage are of even more recent origin, but they are expected to have a major impact on the personal-computer market in the near future. *WORM disks* (the name is an acronym for "Write Once–Read Many") are optical storage devices similar to CD-ROM disks in that they feature permanent recording, but unlike the latter, WORM disks can be prepared by the user. The content of a WORM disk (capacity ca. 500 MBytes) cannot be altered once it has been established, but it remains subject to reading almost indefinitely. WORM disks are thus ideal for archival purposes. *Magneto-optical* (MO) drives are comparable, but they offer more flexibility. The optical disk in this case is coated with a thin layer of magnetic material, which makes it possible through a rather complex read/write process to store (reversibly!) on a single disk 1 GByte or more of data—the equivalent of nearly 1300 double-sided 800-KByte diskettes! *Floptical* drives are relatively inexpensive second-cousins of the traditional 3.5″ diskette drive, but each "diskette" in this case is capable of holding 21 MBytes of information. The principal drawback of flopticals is an access time comparable to that of a diskette. On the other hand, the associated cost per megabyte of information is about half that of the other technologies. Another recent development in this rapidly evolving field is the *dual-function optical drive* capable of accepting both erasable and WORM disks supplied in a special cartridge format.[11]

Mass-storage devices of all types require special file-management software, usually supplied as part of a set of proprietary programs loaded into the system by the manufacturer. Such packages sometimes include convenient

---

[11] For an introduction to optical drives generally see J. Rizzo, "Maximum Movable Megabytes: Erasable Optical Drives" (*MacUser*, November 1990, pp. 102–130). Dual-function drives are discussed by the same author in "Double-Duty Drives: Multifunction Optical Storage", (*MacUser*, November 1991, pp. 108–120). More recent comparative treatments are provided by E. J. Adams in "Storage Strategies" (*Publish*, April 1992, pp. 66–74), M. Baer in "Choosing a Storage Strategy" (*Macworld*, August 1993, pp. 226–121), and J. Rizzo in "Choosing the Right Removable" (*MacUser*, November 1993, pp. 106–112).

backup software (cf. Sec. 5.5), *repair utilities* for dealing with unexpected problems, and *security software* for selective password protection.

All the storage systems discussed in this and the preceding section share a common purpose: facilitating the accumulation in a single place of a wide assortment of programs and documents. Such consolidation can be a great benefit to the user, but it also has the potential for fostering mass confusion, especially in the absence of efficient organization. Even the disciplined user will occasionally confront the problem of a mysteriously "missing" file. Systematically searching a large hard disk arranged according to a hierarchical file system, with files scattered through multiple layers of directories,[12] can be a very time-consuming and frustrating process. A single manual search for a document inadvertently saved to the "wrong" folder (or subdirectory) is usually enough to convince even skeptics of the virtue of acquiring and learning to use a reliable piece of file-location software; examples for the MACINTOSH include Apple's rather primitive FINDFILE DA (Fig. 4–7), DISKTOP (Prairie Software; the user interface for version 3.03 of DISKTOP is shown in Fig. 5–5), MR. FILE (Softways), ON LOCATION (On Technology), and ALKISEEK (Alki Software). XTREE GOLD (XTree) provides similar capabilities to MS-DOS users.

5.5 *Backup Copies, Archival Programs, and Copy Protection*

Diskettes, and especially hard disks, convey a sense of security and permanence that can be quite misleading. Indeed, the *lifetime* of a diskette is likely to be rather limited—extending over a few years at most. Reading and writing data to a disk, especially at frequent intervals, inevitably results in a certain amount of wear on the magnetic storage surface, which ultimately translates into a loss of information. Serious losses are most effectively prevented by disciplining oneself to prepare comprehensive backup copies on a regular basis. Every user should maintain at least one current and complete backup copy of *all* data files, documents, and programs, and no work-

---

[12] It is not unusual for document folders to be "nested" five levels deep!

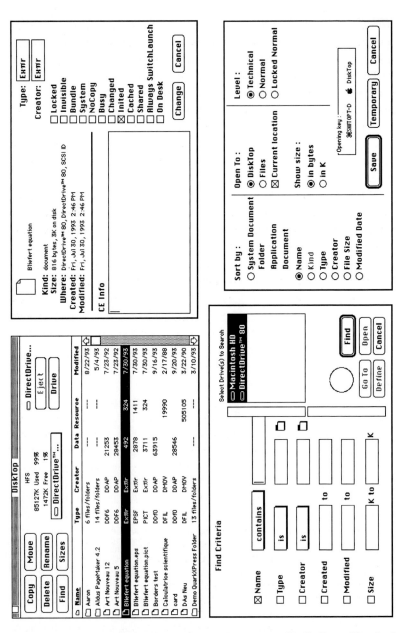

Fig. 5–5. Several screen displays from the MACINTOSH file-finding utility (and desktop replacement) DISKTOP (version 3.0.3).

ing session should be considered complete until backup copies have been prepared of new and revised documents. An even better practice is to make backups several times during the *course* of one's work.

Special *backup software* is available to assist in the rapid preparation of copies of all the files on a given hard disk—or a selected set of files—either to magnetic tape or to a series of diskettes introduced one at a time as required. MACINTOSH programs of this type include Apple's HDBACKUP and the far more versatile DISKFIT DIRECT (Dantz Development) and SNAPBACK (Golden Triangle Computers); FASTBACK PLUS (Fifth Generation) offers similar protection under WINDOWS.

There are definite advantages to ensuring that backup copics arc in the most compact form possible, and to this end various "file compression" utilities have been devised.[13] A simplistic way of visualizing the compression process is to imagine a data file that somewhere contains a repetitive sequence of identical characters—12 adjacent blank spaces, for example. A compression program would deliberately search for such sequences, and then recast them in an abbreviated form. Thus, the 12 blanks of our example might be replaced by the *number* "12", associated in some unambiguous way with the corresponding repeating symbol, in this case the blank. File-compression algorithms are most effective for reducing the size of graphics files containing only black and white data, which can often be compacted to 10–15% of their original size. Even text files are subject to compression by an average of 50%. Accessing a compressed file ordinarily requires that the original process be reversed in the form of a "decompression".[14] Decompression (or expansion) is usually quite rapid, so the extra step constitutes only a very minor nuisance. Some compression routines even provide facilities for automatic decompression of a compressed file whenever an attempt is made to access its content.[15]

---

[13] Many such programs offer the option of also *protecting* the compressed data by limiting access to users familiar with a particular password.

[14] The recently released MACINTOSH program AUTODOUBLER (Fifth Generation Systems) generates compressed files that can in some cases be accessed directly without the need for full expansion.

[15] Examples of such programs for the MACINTOSH include DISKDOUBLER and STUFFIT DELUXE, available at modest cost from Fifth Generation Systems and Aladdin Systems, respectively; WINDOWS and DOS users may wish to consider STACKER (from Stac Electronics). Many of the applications programs distributed today are so large that even they must be released in the form of self-expanding compressed files.

Blank 3.5″ diskettes typically retail for one dollar or less, whereas original program diskettes are often worth several hundred dollars. Consequently, prospective users are occasionally tempted to engage in unauthorized copying rather than paying the retail price for a program. The limits within which one is actually *permitted* to copy commercial software are exceedingly narrow, and they are always spelled out carefully on the sealed envelope that contains the program diskette(s).[16] It should not come as any surprise that software companies have been tempted to seek creative ways of preventing the unauthorized copying of their programs. After all, software "pirating" is thought to cost software developers millions of dollars in revenue every year. Clever programmers have devised a host of sophisticated copy-protection schemes, including

> "extra track methods", "extra sector methods", "dummy sector methods", "spiral track methods", "broad track methods", and "super sector methods"

but equally enterprising entrepreneurs long ago accepted the challenge of developing programs with the sole purpose of *defeating* other programmers' copy-protection schemes. An example from the MACINTOSH world is the legendary COPY-II-MAC, developed, enhanced, and distributed for many years by Central Point Software. In fairness it should be acknowledged that programs of this type serve one very legitimate purpose as well: freeing the user from complications that often accompany an attempt to transfer copy-protected software to a hard disk. Many protected programs installed in the usual way fail to load properly, and others insist that an original diskette always be introduced into an appropriate floppy drive prior to activation of the program from the hard disk (an annoying way of verifying that the user actually *has* such a diskette). Complaints from users, as well as the widespread availability of sophisticated copying programs, eventually resulted in a nearly universal trend away from copy-protection schemes.

---

[16] Guidelines on the subject of "authorized copies" and "pirating" constitute a distinctive branch of national and international copyright law. For recent thoughts on this subject see S. Levy, "The Rap on Software Piracy" (*Macworld*, January 1993, pp. 57–70) as well as P. Roberts, "Electronic Ethics" (*Aldus Magazine*, June 1993, pp. 15–18), which focuses specifically on the copying of graphic material. A more comprehensive source of information, providing an extensive bibliography, has been prepared by the National Research Council (1991).

5.6 RAM Disks, Cache Storage, and Virtual Memory

A *RAM disk* is a transient, fictitious storage "device" conjured up within a computer's own memory: ordinary software is used to redefine a section of RAM as an addressable data-storage register. "Installing" a RAM disk is the process of creating such a reserved memory block, the existence of which is documented through a desktop icon representing the new "disk". Data transfers to and from internal memory occur very rapidly, so running a program from a RAM disk can lead to a dramatic increase in operating efficiency. The major disadvantage, of course, is that this form of "storage" is exceedingly fragile and unreliable: all the stored information will disappear if the computer "crashes", or if there is an unexpected power outage (lasting even an instant). If a RAM disk is used for the storage of *data* (e.g., a text manuscript), it is absolutely essential that a copy of the file be saved *frequently* to a permanent storage medium (a hard disk or a floppy—better yet, both!). A second disadvantage of a RAM disk is that it consumes valuable memory, a luxury many users cannot afford.

RAM disk installation is usually accomplished through a dialog box in which one establishes a set of parameters applicable to the new (virtual) "disk". In principle a RAM disk can be so defined that it serves as temporary housing not only for applications programs and data, but also system files, printer drivers (cf. Sec. 6.1), and the like. Instructions can even be issued for an appropriate RAM disk with a complete set of files to be created automatically each time the computer is restarted.

RAM cache (from the French *cacher*, to hide) represents an interesting variation on the RAM-disk concept. Once again, a portion of the computer memory is set aside for information storage (in the form of the "cache"), but its function this time is to act as a *buffer* between the principal storage medium (e.g., a hard disk) and the main bank of memory. All instructions read from disk are automatically copied simultaneously into the cache. If the same information happens to be requested a second time it can be recovered directly from the cache, a process that consumes considerably less time than conventional disk retrieval. The extent of the cache determines the total number of instructions that can be so retained, and this parameter is established by the user at the time the cache is created (from the MACINTOSH CONTROL

PANEL desk accessory, for example; cf. Fig. 4–5). Under favorable circumstances a RAM cache may eliminate the need for as many as half the program calls that otherwise would be directed to a disk.

Finally, the inverse process is sometimes exploited as well, especially with the latest versions of the MACINTOSH system software: temporary disk storage is in this case used as a way of circumventing a problem of insufficient memory (RAM). If, for example, some operation requires more temporary storage space than is currently available in memory, excess data is simply deposited in a disk file where it can later be consulted as required. In principle this is a convenient crutch for the user unwilling to invest in an adequate amount of memory, but a high price must nonetheless be paid in the form of delays caused by (relatively) long disk-access times. Frequent recourse to virtual memory is a sure sign that a memory upgrade would be warranted.

One other recent and powerful alternative also deserves to be mentioned: an ingenious piece of new software called RAM DOUBLER (Connectix) that invisibly serves to reorganize and more efficiently manage MACINTOSH RAM in such a way that overall memory capacity is increased literally by a factor of two. Thus, a 5-MByte computer is instantly (and inexpensively) transformed into a 10-MByte device with essentially no noticeable impact on systems capability or operating speed. Extensive tests by independent organizations have convincingly verified that use of RAM DOUBLER presents virtually no compatibility problems under normal operating conditions.

6 Output Devices:
Monitors, Printers, and Plotters

6.1 Preliminary Remarks

Electronic manuscripts will never completely eliminate the need for printed manuscripts despite their rapidly increasing popularity. Author-printed copy of varying degrees of quality will continue to find application

- in the editing and revising of text
- as a useful adjunct to an electronic counterpart (in the editorial office, for example)
- as camera-ready copy for professional reproduction

The preferred approach to outputting an electronic document is a function of the circumstances. Screen display is often sufficient during some stages of editing, and a good dot-matrix printer may be perfectly adequate in the case of a routine business letter, but a more sophisticated solution is required for camera-ready copy that will be reproduced for wider distribution.

Regardless of the nature of the task, full compatibility is absolutely essential between the output device of choice and the other hardware and software components constituting a desktop publishing system. In order for a computer to communicate freely with a particular printer it must have ready access to the appropriate *printer driver*: software to guide the operating system in accurately formulating and transmitting the requisite set of printer control messages. The printer driver in effect supplies the information required for transforming print commands issued by other software into signals intelligible to a particular printer, and it is not unusual for a faulty or inappropriate driver to be the cause of apparent compatibility problems. The goal is to transfer the content of some carefully prepared data file—alphanumeric symbols, special characters, and perhaps a few graphics, all suitably prearranged on the basis of a screen display (cf. the WYSIWYG discussion in

Sec. 6.3) to a particular spot on a blank sheet of paper, and in unaltered form.
If unexpected results are obtained, at least some of the troubleshooting attention should be directed toward the software with which the printing device
is managed.[1]

There are many types of printers available for use with personal computers. These can be classified in several ways, including

● as a function of the technique used for transferring ink or an ink-substi-
tute to the printed page

● on the basis of their *resolution*, a common way of expressing the quality
of the finished product (cf. Sec. 6.2)

● in terms of the broad categories *graphic* and *non-graphic*

● according to whether or not they are equipped to interpret a particular
page-description language (e.g. PostScript, Sections 6.6 and 7.6)

Successful output of illustrative material (line drawings, scanned images, etc.)
assumes a printer with *graphics capability*, which means the device is able
to address and selectively interact with precisely defined points on the surface
of the print medium.

6.2 Resolution

Output devices as diverse as display screens *(monitors)*, most printers, and
even high-resolution laser typesetters (*imagesetters*, Sec. 7.5) operate on the
common principle of using appropriate sets of dots to simulate alphanumeric
characters and graphic images that would ordinarily be constructed on the
basis of continuous lines. This is accomplished by selective activation of

---

[1] Just like other software, printer drivers are subject to occasional revision and
"updating" by their developers. In most cases it is best to install the most recent
version of a particular driver. Some programs (e.g., PAGEMAKER) offer proprietary
printer drivers as alternatives to the standard drivers shipped with the printer or the
computer operating system. On rare occasions such special drivers actually interfere
with the output process rather than improving it, as has been documented, for example, with certain PAGEMAKER files containing mathematical formulas prepared
with early versions of EXPRESSIONIST.

specific points within a predefined *matrix* of points. The smaller and more closely arrayed the matrix points, the clearer will be the resulting image, and the more "authentic" it will appear to the eye. High output quality thus implies an exceptionally fine matrix capable of effectively simulating smooth curves, thereby producing reasonably accurate representations of symbols. Figure 6–1 illustrates the principle with two examples, one corresponding to output like that of a typical computer display screen, based on a relatively coarse matrix, and the other showing the same information as expressed by a laser printer.

The image fidelity possible with a particular output device can be treated as a function of the distribution *(point density)* of the primary image-forming elements. It is usually expressed in terms of a *resolution*, quantified according

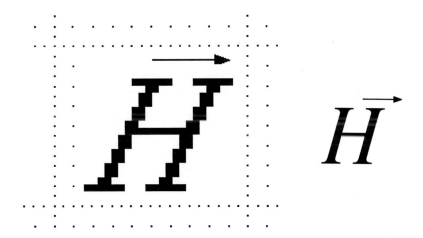

Tabelle 11.3 Datengrundlage für die Einstufu			*Tabelle 11.3*. Datengrundlage für die Einstufu		
Einstufungen	Bei der Berec zu berücksich		Einstufungen	Bei der Berec zu berücksich	
Klasse I			Klasse I		
● Sehr giftig und Giftig	> 0,2 %		● Sehr giftig und Giftig	> 0,2 %	

Fig. 6–1. Screen display (left; 72 dpi) and laser-printer (300 dpi) output of a capital letter surmounted by an arrow (prepared in MACDRAFT), as well as a set of data arranged in tabular form (PAGEMAKER).

to the number of available "dots" per unit of length (occasionally per unit of area); e.g., in *dots per inch (dpi)*.[2] Most output devices provide the same resolution horizontally as vertically, which is why it is more common for a resolution to be expressed as 300 dpi, for example, rather than the equivalent $300 \times 300 = 90\ 000$ dots per *square* inch. Printers, imagesetters, plotters, and display screens all lend themselves to evaluation on the basis of resolution, as illustrated in Fig. 6–2 (cf. also Table 6–1). Generally speaking, the higher the resolution of a particular output device the higher the output quality—and the higher the cost.

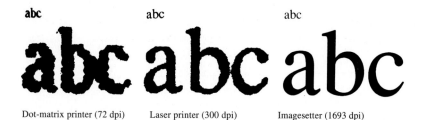

| Dot-matrix printer (72 dpi) | Laser printer (300 dpi) | Imagesetter (1693 dpi) |

Fig. 6–2. Letters produced with printing devices characterized by different levels of resolution (all samples enlarged photographically by a factor of 7).

Table 6–1. Typical resolution ranges for various types of output devices.

Device	Resolution, dpi	Resolution, lines/cm
Monitor	50 ... 300	20 ... 120
Dot-matrix printer	100 ... 360	40 ... 145
Laser printer	300 ... 1200	120 ... 480
Laser imagesetter	> 1000	> 400

---

[2] As the inventors of the metric system, it is understandable that the French would resist use of non-metric units of measure, but in the particular case of resolution even the French rely on dots per inch, translated as «point par pouce» with the abbreviation "ppp". When metric data are preferred for describing resolution, the appropriate unit is *lines* (not dots!) per centimeter.

Less precise qualitative descriptions are sometimes used to categorize printers on the basis of resolution. Thus, a good dot-matrix printer (cf. Sec. 6.4.1) produces output roughly comparable to what one would expect from a typewriter, so it is sometimes described as a "letter-quality" device. Laser-printer output *approaches* that of professional typeset copy, earning it the designation "near-print quality" (cf. Fig. 6–2).

A very simple technique makes it possible (within limits) to enhance the quality of printed output regardless of its source: photographic reproduction at a reduced scale. For example, most of this book was prepared with a high-resolution imagesetter, but certain pages were printed with a laser printer in the form of copy 130 mm wide, permitting subsequent photographic reproduction with a linear reduction of 20% (for more information see Appendix C). The effect is an increase in the apparent resolution of such a page from 300 dpi to 375 dpi, a gain of 25%.[3]

6.3 *Display Screens*

One of the greatest nuisances confronting every desktop publisher is the fact that the resolution of a typical computer display screen falls far short of that achieved with a high-quality printer. The difference is in fact so great that font designers must usually prepare two very different versions of each of their fonts: one rather crude form to serve as a basis for screen characters ("screen-font files") and another highly refined product for hardcopy output ("printer-font files"). This is the case, for example, with all the PostScript fonts (cf. Sec. 7.6) included in the ADOBE TYPE LIBRARY.

Even though text is always displayed on a computer screen in dot-matrix form, this is not the way it is actually stored or manipulated within the computer, in part because matrix-based files are too unwieldy. For example, displaying a single character from the font "Helvetica, 12 point" involves

---

[3] Calculation on the basis of points per unit *area* leads to an even more impressive numerical result. Thus, if the original resolution was 90 000 dots per square inch, the finished product after photoreduction (to 80%) would correspond to 140 625 dots per square inch, an increase of *56%*. Equipment manufacturers obviously prefer this standard when comparing their devices with the competition.

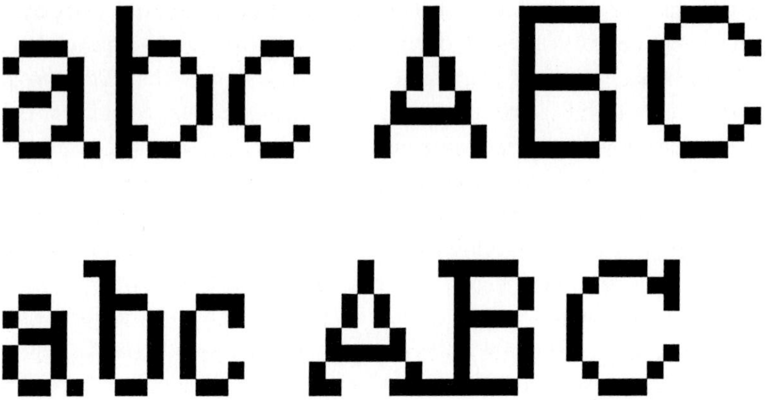

Fig. 6–3. Several 12-point letters (Helvetica, Times) as they would be displayed on a 72-dpi monitor, enlarged to reveal more clearly the underlying bitmap structure.

activating several screen pixels from within a 7×10 pixel matrix, which contains a total of 70 data points (cf. Fig. 6–3). One byte of memory is sufficient for addressing at most 8 display pixels, so approximately 10 bytes would be necessary to define a single letter of the alphabet. A full screen of single-spaced text (e.g., ca. 2500 characters) would therefore require 25 KByte of storage for display purposes alone! For this reason, text manipulation is routinely accomplished with the aid of powerful "macro" commands written in a special graphics language, permitting the alphanumeric information of interest to be communicated more efficiently in terms of its symbolic content (supplemented with appropriate modifiers for establishing such display attributes as size, style, weight, etc.). In this way the information corresponding to any particular character can be condensed into *one* byte of memory (exclusive of special formatting instructions), a savings of nearly 90%.

A delightful array of acronyms has grown out of innumerable attempts to characterize the extent to which text presented on a "typographic display screen" actually resembles its printed counterpart. The one most frequently encountered—albeit perhaps the least applicable—was mentioned in passing in Section 1.2:

WYSIWYG

from "what you see is what you get", an optimistic description ostensibly coined by J. Seybold, editor of the *Seybold Report on Desktop Publishing*. Unfortunately, the claim that a particular system features a "WYSIWYG" working environment often reflects wishful thinking on the part of the developer, valid only in a very crude sense. The point is readily apparent if one compares a MACINTOSH screen display with the corresponding laser-printout, as in Fig. 6–1. Notice particularly in this case the horizontal lines in the table, the position of the letter "G" in the word "Giftig", and the placement of the arrow over the letter "H" at the top of the figure.

Several alternative (if slightly irreverent) descriptive acronyms more nearly reflect the true situation with which the average desktop publisher can expect to be confronted, including

WYSIMOLWYG

("what you see is more or less what you get"),

WYSIPOWYG

("what you see is part of what you get"; Peters, 1988), and

WYSIWHOOPS

("what you see is what happens on operating in PostScript"; cf. Sec. 11.4).

This terminological whimsy underscores in an amusing way the fact that most representations of text and graphics on a computer monitor are seriously deficient, largely due to inadequate screen resolution, and the problem will persist until monitors become available with at least the 300-dpi resolution of a standard laser printer.[4] Equally important will be the development of a reliable and efficient device-independent *screen* description language comparable to the PostScript *page* description language (cf. Sections 6.6 and 7.6).[5]

---

[4] Screens with a resolution of 300 dpi may someday become commonplace in the world of DTP, a welcome prospect if only because of increased screen legibility with respect to small type. A prototype screen with this resolution (the "Exakt 800") was demonstrated by Flanders Research as early as 1988 at the trade show "Type-X" in Orlando, Florida.
[5] Adobe Systems has already developed a provisional form of *Display PostScript* that IBM and others have expressed interest in licensing, a particularly promising development since it permits use of a common approach to controlling both screen and printer. A few computer programs (ADOBE ILLUSTRATOR, for example) already utilize ordinary printer PostScript in their internal screen-output routines.

In the meantime, DTP enthusiasts must reconcile themselves to display screens that devote only 20, 30, or 50 pixels to characters whose laser-printer counterparts are painstakingly assembled from as many as 500 dots. Breuer (1988) has seriously advocated widespread adoption for now of the more realistic acronym

WYSINWYG

("what you see is *nearly* what you get"). So long as screen resolution continues to lag far behind that of printers and phototypesetters the most apt expression of all may be

WYGIWYS

("what you get is what you'll see"). At least for the foreseeable future, desktop publishers will continue wasting significant amounts of time and material in the preparation of one or several *test prints* of each of their carefully assembled pages.

Before leaving the general subject of monitors two other matters warrant at least cursory attention: screen format (and size), and the question of color. The most frequently encountered *format* is still the 14-inch (diagonal measurement) screen, typically 640 pixels wide and 480 pixels high (roughly $8.9'' \times 6.7''$, slightly larger than half of a letter-sized sheet). A more practical device for the writer is the somewhat larger portrait-format monitor consisting of 640×870 pixels, which can produce a full-scale 72-dpi representation of an $8.5'' \times 11''$ letter-sized (or A4) sheet. Better still is the 21" (1152 × 870 pixel) "two-page" display, since it lends itself to the design of two-page spreads.[6]

From an aesthetic standpoint there are certainly benefits to be derived from access to a good *color* monitor, but the average writer would be hard-pressed to make a strong case for acquiring color capability. After all, the finished product of an author's efforts will almost always be "black and white". Moreover, a high-quality *monochrome* monitor typically provides a significantly sharper image than a color monitor, and experience suggests that such "convenience features" as multicolored highlighting add little to the creative

---

[6]   A large monitor is certainly convenient, but its role in the creative process should not be overemphasized. This book (like many others) was prepared quite comfortably with a "compact" MACINTOSH SE/30 computer equipped only with a standard 9" MACINTOSH internal monitor (viewing window: $7'' \times 5''$).

process. Color also has the distinct disadvantage of forcing the computer to work markedly more slowly. The same arguments apply (though less forcefully) to *gray-scale* monitors, which offer perhaps 256 shades of true gray rather than the millions of hues available with a color monitor. Unless one is planning to work extensively with halftone images or complex color illustrations it is probably wiser to spend the extra money required for a gray-scale or color monitor on additional memory or hard-disk capacity.[7]

6.4 Printers and Plotters

6.4.1 Dot-Matrix Printers

Matrix printers are printing devices that simulate letters, line drawings, and continuous-tone images with intricate patterns constructed from tiny dots, where each dot is a discrete member of an extensive *matrix* of dots (cf. Sec. 6.2). This printing technique is the basis not only of the relatively low-resolution mechanical devices traditionally known as "dot-matrix printers" but also of ink-jet and laser printers, which differ from the former only in the fact that they provide access to much more closely-spaced dots. Dot density is the primary factor in determining the output quality of any matrix-printing device, just as with a monitor.

Dot-matrix printers in the traditional sense of the term rely on mechanical impact to transfer dots of ink from a print ribbon to a sheet of paper. For many years these were the preferred hardcopy output devices for personal computers because of their low cost, relatively high speed,[8] technical simplicity, and versatility. In contrast to laser printers (Chap. 7), mechanical dot-matrix printers are usually designed to accept not only single sheets of paper but also tractor-fed, fan-folded media, and they are uniquely applicable to printing on multipart forms and spirit duplication masters ("ditto masters"). The general operating principle of such a device is straightforward: an electronic signal from the computer is interpreted within the printer and then

---

[7] For a recent discussion of the subject see R. C. Eckhardt, "Here's Looking at Monochrome Monitors" (*Macworld*, May 1993, pp. 113–119).
[8] The *speed* of a mechanical printer is expressed either in CPS (characters per second) or LPM (lines per minute).

passed along to a *print head* outfitted with an array of tiny, movable pins (typically 7 to 48 in number depending on the quality of the printer; the original Apple IMAGEWRITER, for example, was a 9-pin device). The pins themselves are usually arranged in a regular pattern of rows and columns, with certain columns perhaps offset slightly relative to others. Producing a single line of text requires that the print head execute one or several complete passes across the width of the paper, in the course of which tiny hammers are caused to strike specific pins at precisely defined intervals, thereby transferring minute dots of ink to the paper from a ribbon located between the paper and the pins. The resulting dots are so positioned that they form a complex pattern closely resembling a line of fully formed letters. Depending on the number of available pins, output from a printer of this type may warrant the designation "letter quality" (24 pins or more; cf. the 27-pin Apple IMAGEWRITER LQ) or, with fewer pins, "near-letter quality".

The least expensive models rely on fewer than 10 pins, and they are responsible for the notoriously poor copy associated with the pejorative term "computer printout". Clearly such a printer has no meaningful role to play in desktop publishing, although somewhat better results are sometimes achieved with these devices by increasing the number of print-head passes devoted to each printed line, with the dots from one pass offset slightly relative to those from the preceding pass (cf. Fig. 6–4). Print resolution can be increased in this way by as much as a factor of two (e.g., from 72 dpi to 144 dpi). Printing speed can be increased by providing for *bidirectional printing*, in which the first pass for a given line is printed from left to right and the second from right to left as the print head returns to its initial position. Some printers offer the choice of either unidirectional or bidirectional printing.

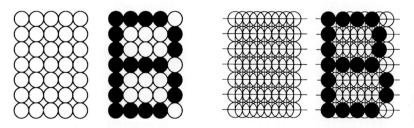

Fig. 6–4. The letter "B" as represented within a 5 × 7 dot matrix, showing the improvement possible with two-pass printing.

6.4.2 Ink-Jet, Daisywheel, and Thermal Printers

Ink-jet printers are somewhat analogous to laser printers (Chap. 7) in that both are matrix printers of the *non-impact* type. An ink-jet printer uses microscopic nozzles to spray liquid or semi-solid ink onto the print medium from a disposable cartridge. Some models eject ink directly in the form of discrete droplets, but more advanced devices are designed to release their ink in a continuous stream that is subsequently broken into microscopic droplets of a consistent size by an ultrasonic generator located just above the surface of the paper. Ink can sometimes be delivered in this way at a rate of over 100 000 droplets per second. The stream of microdroplets is caused to pass near a charged electrode, from which each droplet acquires an electrostatic charge. Variations in the level of charge make it possible to use a pair of *deflector electrodes* to establish precise trajectories for individual droplets, so that each is directed to a precisely determined location on the page. Once the ink droplets reach the paper they have some tendency to spread and flow together, leading to images with somewhat smoother contours relative to other matrix printing devices.[9]

Ink-jet printers are characterized by fairly rapid throughput (e.g., two pages per minute for text), moderate cost, relatively high resolution (typically 300 dpi or greater), and low noise. They are also capable of reproducing graphic images— including halftones—but the process is often quite time-consuming, especially when compared with a laser printer (cf. Chap. 7). Prices for ink-jet printers have fallen dramatically in recent months, with the result that laser-quality output is now available from an ink-jet device costing little more than a good mechanical dot-matrix printer. Ink-jet printers also open the way to inexpensive high-resolution color printing, although color copies take even longer to prepare. Finally, it should be noted that, while the printer itself may be inexpensive, *printing* with an ink-jet printer can be more costly than laser printing, because the ink cartridges are rather expensive, and special paper is required for optimal output.

---

[9] Ink jet printers also tend to drive the ink into the paper fibers to such an extent that the print becomes almost impervious to mechanical abrasion. By contrast, toner from a laser printer is essentially "ironed on" the paper surface (cf. Sec. 7.2), so most of it remains subject to removal by careful scratching with a sharp knife. Some inks used with ink-jet printers are water soluble, causing them to smear if the paper becomes wet.

Daisywheel printers represent a mechanical alternative for preparing high-quality printed text. Like a typewriter, a daisywheel (or type-ball) printer takes advantage of a print head that features a full set of carefully crafted typographic characters. To produce the various printed characters the print head is repositioned as required while the carriage transporting it passes over the surface of the paper; ink is transferred to the page from a ribbon, just as with a typewriter. While the print quality achievable with such a device is excellent, the only way to change the type size or type style is to replace the daisywheel itself, and no facilities are provided for reproducing graphics. Like dot-matrix printers, daisywheel printers have been largely displaced as prices for more versatile printers have declined.

A *thermal printer* operates on the basis of a print head equipped with a matrix of tiny resistors subject to selective heating electrically to a temperature of ca. 120 °C. Printing in this case requires use of a special *thermal paper* that darkens locally upon exposure to heat. Many facsimile printers ("FAX machines") operate on this principle. *Thermal transfer printers*, by contrast, use special ribbons from which heated dye can be transferred to ordinary paper. Thermal printers of both types are suitable for graphic output, and they offer high resolution and quiet operation, but the advantages are largely offset by the high cost of the requisite special paper (which also has a limited lifetime) or special ribbons, and the fact that the performance of thermal print elements is subject to degradation over time.

6.4.3 Plotters

Plotters (also known as *xy-plotters*, *digital plotters*, or *drafting machines*) are output devices specifically created for the accurate reproduction of curves expressed in mathematical form. Two basic designs are relatively common: *flatbed plotters* and *drum plotters*. With a *flatbed* plotter the paper is held firmly in place on a planar surface, and the drawing device (which is equipped either with high-quality drafting pens or disposable fiber pens) is moved across the paper surface under the impetus of two computer-driven motors, one for the *x*- and one for the *y*-direction. Directing appropriate impulses simultaneously to the two motors makes it possible to construct not only horizontal and vertical lines but also lines that run diagonally, or accurately follow curved paths. Additional mechanical linkages are present for raising the pens and lowering them again onto the paper. The drawing device of a *drum* plotter moves only in a single direction: back and forth across the width of the paper. The paper is wrapped around a metal drum that is in turn subject to precise rotation in the backward and forward directions.[10]

Whereas an image created with a matrix printer always consists of a collection of individual dots, plotter output is based on true lines, which can be caused to have virtually any desired absolute width. Plotters are thus especially well suited to the preparation of large-scale precision line drawings, such as building plans or maps, and the quality of the corresponding output is exceptionally high. Many plotters are equipped with facilities for several pens, permitting the preparation of multicolored drawings.

Plotters, like printers, require specialized driver software to translate digital data from a computer program into commands for controlling the various motors. The program MACPLOT (from Microspot) provides all the software tools required for reproducing on plotter paper or vellum a graphic image prepared originally with a standard graphics application such as DELTAGRAPH PROFESSIONAL (DeltaPoint; cf. Sec. 12.3) or MACDRAW (Claris; cf. Sec. 11.3). Support is included for multiple colors as well as complex fill patterns. A plotter is ideal for preparing technical drawings of all types to a high degree of accuracy and at any desired scale—and the results are entirely free of the "steps" inevitably present in diagonal or curved lines produced with a matrix-type device—including a laser printer, as close inspection will quickly reveal.

The MACPLOT software mentioned above is compatible with more than 20 different commercial plotters (cf. Fig. 6 5), and generic drivers are included for dealing with more obscure models. A simple "copy" command suffices to transfer graphic data to the MACPLOT clipboard (an intermediate storage facility), which is represented by one of three windows that appear when the program is loaded. A second window is used for selecting pens appropriate to the various lines in the drawing, and for assigning MACPLOT fill patterns. The third is for configuring the actual plot operation, where the user is free to specify the precise position at which some particular graphic element is to appear on the overall drawing surface. The precision with which data can be manipulated with MACPLOT is suggested by the fact that lines constituting a fill pattern can be separated to any extent desired in the range 0.02–10 mm!

---

[10] Both of the plotter types described here could be classed as *mechanical*. An alternative is the *electronic* plotter, which dispenses with all moving parts. In this case a computer-guided cathode ray is used to generate a primary image on a cathode-ray tube, and this electronic image is then captured on photographic film or photosensitive paper.

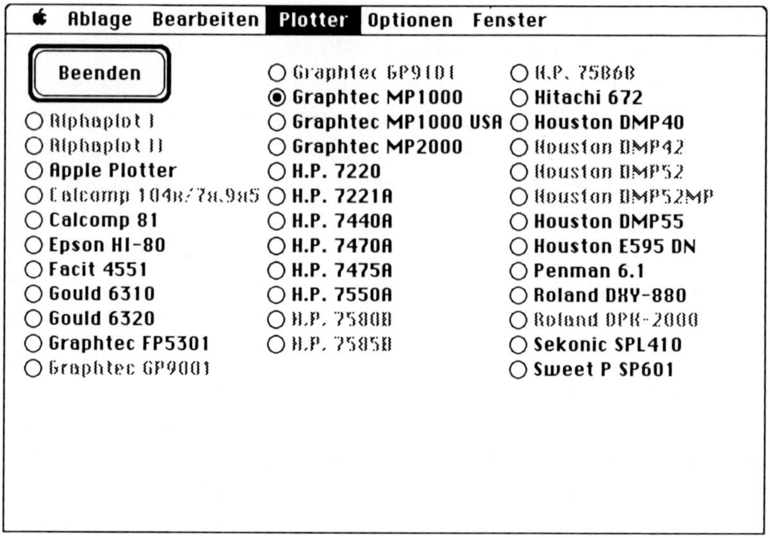

⚫ Ablage	Bearbeiten	Plotter	Optionen	Fenster

Beenden	○ Graphtec GP9101 ○ H.P. 7586B
	⦿ **Graphtec MP1000** ○ **Hitachi 672**
○ Alphaplot I	○ **Graphtec MP1000 USA** ○ **Houston DMP40**
○ Alphaplot II	○ **Graphtec MP2000** ○ Houston DMP42
○ **Apple Plotter**	○ **H.P. 7220** ○ Houston DMP52
○ Calcomp 104x/7x,9x5	○ **H.P. 7221A** ○ Houston DMP52MP
○ **Calcomp 81**	○ **H.P. 7440A** ○ **Houston DMP55**
○ **Epson HI-80**	○ **H.P. 7470A** ○ **Houston E595 DN**
○ **Facit 4551**	○ **H.P. 7475A** ○ **Penman 6.1**
○ **Gould 6310**	○ **H.P. 7550A** ○ **Roland DXY-880**
○ **Gould 6320**	○ H.P. 7580B ○ Roland DPX-2000
○ **Graphtec FP5301**	○ H.P. 7585B ○ **Sekonic SPL410**
○ Graphtec GP9001	○ **Sweet P SP601**

Fig. 6–5. Dialog box from the program MacPlot (German version) for specifying graphic output to a particular type of plotter.

6.5 Print Spoolers

Printing a single page of output containing complex graphic data may consume as much as several minutes even with a laser printer—time that most users would prefer to invest in other activities. Printing via a network can also prove to be a time-consuming and frustrating experience, especially if the network includes several active users. Unprofitable delays associated with printing can be reduced significantly by taking advantage of a *print spooler* (where "SPOOL" is the acronym for "Simultaneous Peripheral Operations On Line"). A spooler is a facility designed to intercept data otherwise destined for a printer (PostScript commands, for example), causing the information to be stored temporarily either in memory or on a disk. Captured information is later released to an available printer at a rate just sufficient to keep the printer occupied, leaving the user (and the computer) free to pursue other useful tasks. In the case of network operation this provides each user with a sense of "private" (albeit delayed) access to a single printer. An alternative term frequently associated with such a process is *background printing*.

Print spooling can be accomplished on the basis of either software or hardware. *Software spoolers* arrange for short-term storage of incoming print data within the context of existing computer facilities (RAM, free disk space), simultaneously maintaining continuous contact with the printer as a way of monitoring its status. The computer operating system plays virtually no part in the process, permitting it to engage itself in other unrelated activities. A *hardware spooler* is somewhat more complicated, in that it provides its own storage facility and utilizes its own microprocessor for information exchange with the printer.

The MACINTOSH operating system is equipped with its own background print utility, but more versatile software spoolers are available from other sources. A dialog box used for establishing spooling preferences with one such spooler, SUPERLASERSPOOL (Fifth Generation Systems), is illustrated in Fig. 6–6.

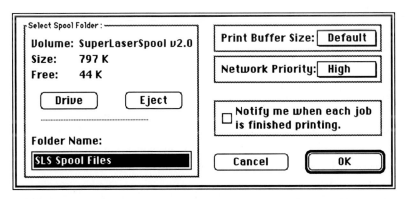

Fig. 6–6. Dialog box from the MACINTOSH print-spool program SUPERLASERSPOOL.

6.6 Page-Description Languages

One conceivable approach to providing a printer with the information necessary for generating a page of text and graphic output would be to transmit on a point-by-point basis an appropriate matrix of dot values specifying either black or white output—or even a wide range of colors. Indeed, some dot-matrix printers (cf. Sec. 6.4.1) are actually designed to be addressed in this way, but the system is clearly impractical for high-resolution output: describing a single letter-size page in this way at a resolution of 300 dpi would entail the temporary storage and communication of data associated with more than eight million points. Moreover, a given matrix is applicable only to one particular resolution, and the associated coding must necessarily be tailored exclusively to a particular type of printer. Device-specific data output of this type is sometimes referred to as "low-level page description".

A more economical, device-independent strategy employs instead a "problem-oriented" description of the document in question. This is accomplished most effectively with the aid of a *page-description language (PDL)*, such as POSTSCRIPT (developed by Adobe Systems), INTERPRESS (Xerox Computer Services), or DDL (Imagen). The general nature of such a language should become apparent from the brief discussion in Sec. 7.6 of the most widely used PDL, POSTSCRIPT.

7 Laser Printers and Laser Typesetters

7.1 Introduction

As noted in Chapter 6, *laser printers* fall within the general category of non-impact matrix printers. All the characters and images produced are actually composed of discrete dots, though they are not usually so perceived by the eye. The fact that a laser printer is capable of creating exceptionally large numbers of tiny dots is responsible for their correspondingly high resolution (cf. Sec. 6.2). Like ink-jet printers—and in contrast to impact-type dot-matrix and daisywheel printers—laser printers operate with relatively little noise. Laser printers are also much faster than daisywheel (or type-ball) printers, which are usually limited to about 90 characters per second.

The actual throughput one can anticipate with a laser printer depends upon

- the amount of text associated with a typical page,
- the number of typefaces and type styles required, and, above all,
- the number and nature of any graphic elements.

Printing one page of single-spaced 12-point Times with a "personal" laser printer (e.g., an Apple PERSONAL LASERWRITER NTR, or a Hewlett-Packard LASERJET IIIP), for example, typically entails about 15 s, whereas a more powerful laser printer (Apple PRO 600, H-P 4M) might complete the same job in half that amount of time—and at twice the resolution! By contrast, several minutes could elapse before a page of complicated graphics would emerge.[1]

---

[1] The relatively long time required to transform graphic elements into appropriate sets of dots is one of the greatest weaknesses of the PostScript page-description language (cf. Sec. 7.6).

Access to a laser printing device is absolutely essential for anyone committed to fast, high-quality output of scientific or technical documents.[2] The most important prerequisite to remember in selecting a particular printer is full compatibility with one's computing facilities, and any proposed combination should be subjected to a rigorous set of direct tests. Other considerations include the precise role that the printer will be expected to play and the level of perfection sought. Among the technical points worth investigating are the following:[3]

- print rate (the average delay in seconds prior to appearance of the first page, and the time required to produce 10 identical text pages as well as 10 pages that differ)[4]
- resolution (in dpi)
- PostScript compatibility (see below)
- maximum paper size: letter (8.5″ × 11″) legal (8.5″ × 14″), or tabloid (11″ × 17″)
- facilities for multiple paper sizes, envelopes, and single-sheet insertion
- number and nature of resident character sets, and the extent of each
- maximum print width (mm or in.)
- buffer capacity (KByte)
- SCSI-port provisions for adding a hard drive
- toner-cartridge characteristics (availability; price; lifetime in pages; potential for refilling)
- number and nature of data-communication ports; manual vs. automatic mode switching (e.g., MACINTOSH–IBM)

---

[2] In the words of E. Ulrich [*CHIP* **1987**, 5,86]: "Desktop publishing—the term itself was once synonymous with the combination of a MACINTOSH and a LASERWRITER. Those days are now past, because layout and typography programs have become available for almost every computer, but that unfortunately increases the risk that desktop publishing will acquire a bad name. One should never forget that the quality achievable with a printed work depends not at all (or only slightly) on the software used in its preparation; the crucial factor is the quality of the associated printer. To put it another way: a cheap computer driving a sophisticated laser printer is far more supportive of desktop publishing than an expensive program limited by a miserable dot-matrix printer."
[3] This list has been adapted from a set of suggestions in an article on word processing in a special issue (1987) of *Computer Persönlich.*
[4] Print rates are usually expressed in pages/min, but a distinction must usually made between the print rate with *variable text* and that for *copy mode* (one-time translation of a page into a dot pattern, followed by multiple reproduction).

- available accessories
- noise level (dB)
- dimensions
- weight

7.2 The Principles Underlying Laser-Printer Operation

The laser printer is just one example of the broader category of *electrographic printers*. Early laser printers were all based on a single internal print mechanism, developed and manufactured by the Japanese firm Canon. Engineers at Canon capitalized on their experience with photocopying equipment to devise a technology so cost-effective that even today many printer manufacturers obtain their basic print engines from Canon, subsequently embedding them in proprietary electronic shells. Canon referred to its original mechanism as an "LBP unit" (for Laser Beam Printer), in which all the crucial elements were combined in a single removable cartridge. The cartridge itself is good for perhaps 2500–4500 prints, after which it can simply be discarded—a technological solution that effectively circumvents potential servicing problems, albeit in an environmentally questionable way.[5]

A laser printer operates on the same principle as a xerographic copier, the only difference being that the primary dark and light signals come not from a printed original but from a laser beam modulated by a stream of computer data. Thus, the computer issues a series of coded messages *describing* a document, one page at a time, typically expressed in the special language known as *PostScript* (see Sec. 7.6 for further discussion of the language itself). A small data-processing device within the printer transforms the signals as they arrive into the equivalent of a bitmap representation, at the appropriate resolution, of all the various elements constituting the current page (letters, symbols, illustrations, etc.).[6] Guided by a mirror, a

---

[5] Cartridge suppliers have begun to respond to the environmental concern by urging that used cartridges be returned (without cost) for central disposal.
[6] With a non-PostScript printer, information arriving from the computer would already be in bitmap form.

modulated laser beam transfers the resulting bitmap image—one line at a time—to the metallic surface of an electrically charged, light-sensitive drum. With each burst of light a tiny portion of the drum surface loses its electrical charge, producing an invisible record of the bitmap in the form of a microscopic pattern of charged and uncharged zones on a metallic surface. In a subsequent step, a colored substance *(toner)* is brushed over the selectively charged surface. The toner consists of a powdered synthetic dye that also carries an electrostatic charge—in this case positive—causing it to cling to those parts of the drum that have previously been discharged, and leading finally to a visible image (albeit only on the drum surface). Next, the drum is brought into contact with a sheet of negatively charged paper, which strips the toner particles from the drum. In a final step, the paper with its selective dusting of toner particles is passed between a pair of heated rollers, causing the toner to be bonded thermally (at ca. 200 °C) to the paper surface. As with a photocopier, the printed sheet is ultimately released into a receiving tray, after which the drum is cleaned and recharged in preparation for the next cycle (see the schematic illustration in Fig. 7–1).

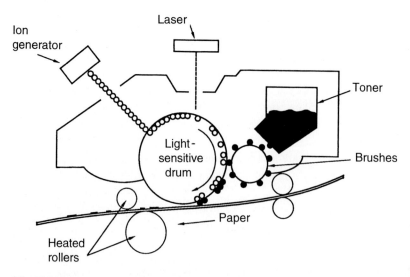

Fig. 7–1. Schematic diagram of a laser printer.

The relative merits of a particular laser printer have little to do with the actual printing mechanism—all printers are based on essentially the same type of device. Much more important is the control mechanism, known as the *raster image processor* (RIP). The RIP of a PostScript laser printer is in effect a PostScript *interpreter*: it translates a page description expressed in the PostScript language into a bitmap (dot-matrix) image, pixel by pixel, creating a structure consistent with the maximum resolution of that particular printer. The RIP is most often a part of the printer itself, permanently embedded in the printer's own microprocessor ("hardware RIP"), but it could also take the form of software for accomplishing the necessary mathematical transformations within the host computer ("software RIP").

Constructing an appropriate bitmap is the most time-consuming part of a PostScript printing operation, so overall output rate is a sensitive function of a particular printer's processing speed and memory capacity. For example, the image depicted in Fig. 11–9 took slightly more than 2 min to print with an APPLE LASERWRITER II NTX (2 MByte of RAM), but 14 min with one of its predecessors, the LaserWriter Plus (1.5 MByte of RAM), even though both printers rely on the same type of microprocessor and operate at the same resolution: 300 dpi. By contrast, a Linotype laser imagesetter (resolution 1694 dpi; cf. Sec. 7.5) required 35 minutes to complete the same task.

Documents prepared with a laser printer convey an impression of overall quality and sharpness approaching that of photo-offset material ("near-print quality"; cf. Sec. 6.2), a feat that until recently could be achieved only with a far more elaborate and costly operation. The breakthrough can be attributed largely to the fact that a laser beam can be focused with very great precision, permitting a standard laser printer (with a resolution of 300 dpi) to work with nearly 8 million dots in the course of preparing a single letter-sized page.[7]

LCS (liquid-crystal shutter) printers represent an interesting alternative to true laser printers, particularly since they offer comparable resolution while

---

[7] The standard for laser-printer resolution has always been 300 dpi, but higher-resolution laser printers are becoming increasingly common (e.g., the Apple LASERWRITER 360 and PRO 600, Hewlett-Packard LASERJET 4M, and VARITYPER VT 5100E, all rated at 600 dpi, or the GCC SELECTPRESS 1200 and the RAM-enhanced Xante ACCEL-A-WRITER 8100, boasting 1200 dpi plain-paper printing). Even 300-dpi *color* laser printers have now become a reality, from companies like Canon, GCC, NEC, Océ, Seiko, and Tektronix (see, for example, the "1992 MacUser Color Buyer's Guide" in the May 1992 issue of *MacUser*).

occupying less space. Until recently LCS printers were also less costly to acquire and operate. The underlying principle is actually not very different from that of a laser printer. The light source in this case is a tubular halogen lamp rather than a laser beam. Mounted between the lamp and the light-sensitive drum is a kind of "ruler" consisting of a linear array of tiny "liquid-crystal shutters". The light transmission characteristics of each of each of these "shutters" is a function of its polarity and the way it responds to an electric field. If this field is so adjusted that it matches the desired print pattern, light will be permitted to reach the drum only at the appropriate points. Thus, a complete "electrical image" is once again created on the drum surface, this time as the drum slowly rotates, one "line" at a time, past the row of adjustable shutters. Beyond this point the two technologies are essentially identical.

7.3 PostScript-Compatibility and Laser Typefaces

With a low-end laser printer like the basic HP-LASERJET,[8] an appropriate bitmap must be retrieved from the printer's own memory for every requested alphanumeric character—in the desired typeface and type size. Special provisions are required for every font one proposes to utilize, and in all its various sizes, which obviously implies the need for a considerable amount of memory.

PostScript-compatible printers, on the other hand, operate on an entirely different principle. Text in this case is constructed from *mathematical descriptions* of *outlines* corresponding to the various alphanumeric characters. Every character in a particular typeface (the Times italic z, for example) is represented by its own unique mathematical formulation (based in turn on Bézier curves), from which a bitmap for that character can be prepared as required in any size desired (for more information see Sec. 7.6).

---

[8]   Because of its wide acceptance the HP-LASERJET has come to be regarded as something of a standard, especially well-suited to the preparation of such straightforward printed matter as business forms and internal memos. The most important consideration in selecting a non-PostScript laser printer is ensuring access to a sufficiently wide array of type families and sizes.

Until quite recently PostScript laser printers were invariably quite expensive, due in part to license fees levied by the developer of the PostScript language (Adobe Systems). Several companies attempted to circumvent the problem by writing their own PostScript-like interpreters, or PostScript compatible "work-arounds", and with such success that prices for true PostScript capability have now fallen dramatically. The Apple MACINTOSH line of computers has since its inception been especially closely associated with the PostScript output system, and PostScript resides at the heart of most of the printers in the Apple LASERWRITER series.[9]

The ROM of a PostScript laser printer typically incorporates not only an interpreter for the PostScript language but also a limited library of PostScript typefaces. The first LASERWRITERS provided direct access to only four character sets: the proportional faces *Helvetica* and *Times*, the monospaced "typewriter" face *Courier*, and the "pi" font *Symbol* (cf. Fig. 2–10 and Sec. 3.9), all but the latter in the style variants plain (roman), italic, bold, and bold italic (other styles such as "underlined" or "shadow" are created artificially). Each of these typefaces thereby became instantly accessible for use with virtually any application program or document—provided a screen version had also been installed in the computer's operating system. Several more fonts were later added to the basic repertoire, including:

Avant Garde
Bookman
Helvetica Condensed
New Century Schoolbook
Palatino
Zapf Chancery
Zapf Dingbats (cf. Fig. 2–9)

A bewilderingly wide variety of typefaces from innumerable sources is now available in the form of *downloadable files* for temporary installation in PostScript-compatible printers and phototypesetting devices. Down-

---

[9]   The license fees even encouraged Apple itself to announce plans to develop (in conjunction with Microsoft) a new page-description language (TrueImage), based directly on QuickDraw routines. Shortly thereafter, Adobe changed its marketing strategy and opened the way to relatively unrestricted access to PostScript technology. For a recent comprehensive review of currently available PostScript laser printers at all levels of sophistication see B. Fraser, "Choosing the Right Laser Printer" (*MacUser*, September 1993, pp. 124–164).

loadable fonts range from the classic to the exotic, and most desktop publishers avail themselves of the opportunity to expand their font collections in this way. Even non-PostScript devices are now able to reproduce PostScript fonts thanks to the program ADOBE TYPE MANAGER (ATM), which performs the necessary conversions to bitmap data within the host computer.

Before leaving the subject of typefaces, brief mention should be made of one important alternative to PostScript fonts: *TrueType*. TrueType represents a different approach to mathematical character description, one that is based on (quadratic) *B-spline* curves rather than (cubic) *Bézier* curves. More important to the user, TrueType characters can easily be interpreted and transformed into bitmaps by a host computer, requiring no intervention on the part of a printer-resident processor. This permits relatively rapid, high-quality text output with *any* printer (or screen display), so long as the appropriate TrueType font descriptions are available to the computer. TrueType capability has recently become a standard part of the MACINTOSH operating system, and it is also the basis for text output under WINDOWS, one of the reasons why the range of available TrueType fonts expanded very rapidly after initial introduction of the technology in 1991. Qualitatively there appears to be little difference between PostScript and TrueType output, although TrueType fonts tend to produce slightly more legible text at small sizes (e.g., 4–8 points). Nevertheless, it is by no means certain that TrueType will displace the venerable PostScript standard, in part because of the heavy investments already made in PostScript by service bureaus and individual users,[10] but also because interpreting the TrueType character descriptions requires a significant amount of computer time, resulting in slower output. On the other hand, if one is constrained to working with a high-resolution printer that lacks PostScript capability it would be wise to investigate the TrueType alternative—and to compare it, for example, with the performance of the alternative ATM approach to PostScript fonts. One final observation: one should exercise considerable caution in attempting to mix PostScript and TrueType fonts indiscriminately within a single document, since the results can sometimes be unpredictable.[11]

---

[10] Note, however, that software has also been developed for converting PostScript (Type 1) fonts into their TrueType equivalents, including the two Altsys programs METAMORPHOSIS and FONTOGRAPHER and Ares Software's FONTMONGER.

[11] For more information on the relative merits of TrueType vs. PostScript fonts see K. Tinkel, "Dueling Font Standards" (*MacUser*, October 1991, pp. 165–177) and J. Felici, "Postscript Versus TrueType" (*Macworld*, September 1991, pp. 195–201).

7.4 Printing with a Laser Printer

Unlike an ink-jet printer, a laser printer imposes very few limitations with respect to the nature and quality of the print medium. Ordinary 20-lb. copier paper produces very acceptable results, but specialty papers of many types are available as well, including ones designed specifically to produce optimum results with laser printers. *Coated* papers offer the advantage of being very smooth and exceptionally white, and they are usually preferred for camera-ready copy (cf. Sec. 2.13).[12]

Paper is normally introduced into the printer from a special *cassette* of precisely the right size. Besides being convenient this ensures that each sheet will be properly aligned, although single-sheet feed is a valuable option if one wishes occasionally to print on overhead transparency film, envelopes, or unusually heavy stock.

An often overlooked advantage of laser printers is the fact that they can easily reproduce a document at any desired *page size*. Precise enlargement up to 400%, or reduction to as little as 25%, is accomplished simply by entering the appropriate parameter into a screen dialog box in the course of requesting a print. It is important to remember, however, that print cannot normally be extended all the way to the edge of a letter-sized sheet. The maximum print area with an Apple LASERWRITER is ca. 206 mm × 269 mm (8.11″ × 10.61″).

Printing to a LaserWriter or similar printer from within a typical MACINTOSH application program begins with a "pulldown menu" command labeled "Page Setup…", which results in a dialog box like the one shown in Fig. 7–2. Here one establishes fundamental parameters that define the character of the proposed document:

- paper size (choice of several options)
- extent of enlargement or reduction
- page orientation ("portrait" vs. "landscape")

---

[12] A particularly wide assortment of laser papers is offered through the catalog of PaperDirect (Lyndhurst, NJ).

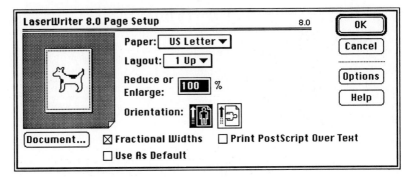

Fig. 7–2. MACINTOSH dialog box for establishing basic output parameters with a LASERWRITER.

- desirability of various special treatments, including
  - font substitution (automatic replacement of any bitmapped fonts in the document by laser fonts; e.g., *Geneva* → *Helvetica, New York* → *Times*);
  - text and/or graphics smoothing (calling upon special algorithms that attempt to improve the appearance of bitmapped images through mathematical interpolation);
  - "faster bitmap printing" (causing certain image-processing steps to be carried out within the computer)

An "Options" switch leads to a second menu (Fig. 7–3) providing the opportunity to

- specify either of two "mirror-image" forms of the document
- reverse the document's tonalities (black → white, etc.)
- approach more closely than normal the edges of the paper
- process more efficiently a document that depends upon a multitude of non-standard (downloadable) fonts

The actual printing process is initiated from a separate "Print ..." dialog box, as shown in Fig. 7–4. Here one normally specifies the number of copies required, the page range of interest, whether or not a separate "cover sheet" should be prepared, and the source of the print medium (cassette or single-sheet feed).

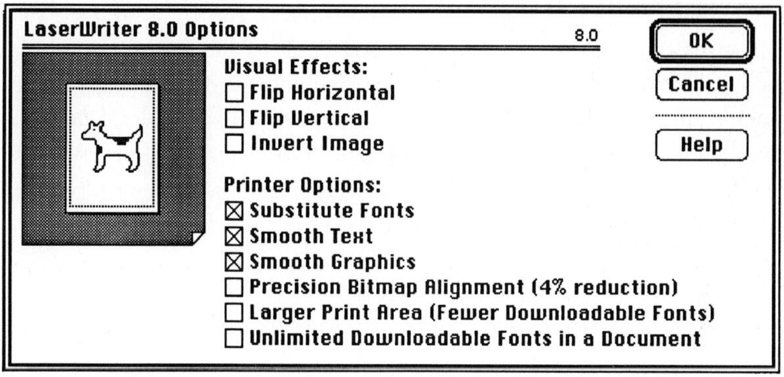

LaserWriter 8.0 Options   8.0

OK

Cancel

Help

Visual Effects:
☐ Flip Horizontal
☐ Flip Vertical
☐ Invert Image

Printer Options:
☒ Substitute Fonts
☒ Smooth Text
☒ Smooth Graphics
☐ Precision Bitmap Alignment (4% reduction)
☐ Larger Print Area (Fewer Downloadable Fonts)
☐ Unlimited Downloadable Fonts in a Document

Fig. 7–3. Additional output "Options" available from the dialog box shown in Fig. 7–2.

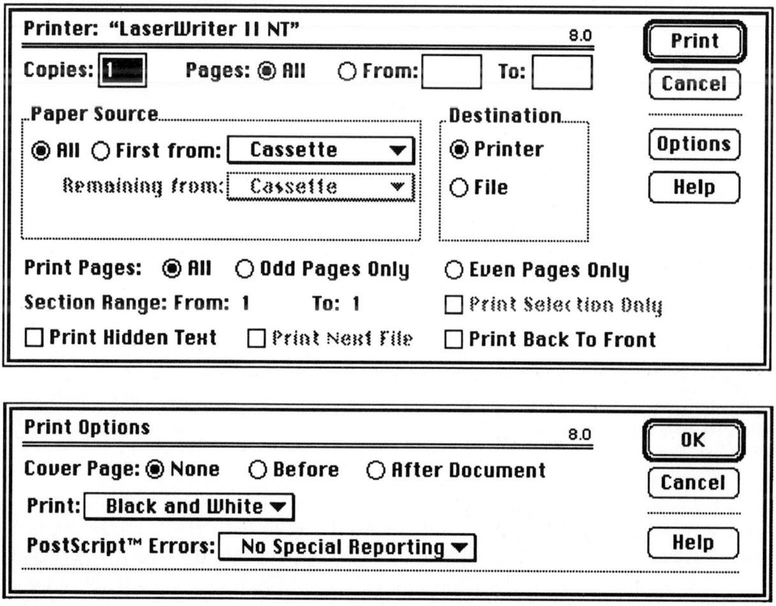

Printer: "LaserWriter II NT"   8.0

Print

Cancel

Options

Help

Copies: [1]   **Pages:** ◉ All   ○ From: [ ]   To: [ ]

Paper Source
◉ All ○ First from: [Cassette ▼]
      Remaining from: [Cassette ▼]

Destination
◉ Printer
○ File

Print Pages: ◉ All   ○ Odd Pages Only   ○ Even Pages Only
Section Range: From: 1   To: 1   ☐ Print Selection Only
☐ Print Hidden Text   ☐ Print Next File   ☐ Print Back To Front

Print Options   8.0

OK

Cancel

Help

Cover Page: ◉ None   ○ Before   ○ After Document
Print: [Black and White ▼]
PostScript™ Errors: [No Special Reporting ▼]

Fig. 7–4. MACINTOSH dialog box for initiating the actual printing process with a LASERWRITER.

Computer-driven laser printers have become so readily available in recent years that they are often called upon in circumstances that would be served equally well by a typewriter. This is understandable given the high quality of the printed product and the convenience of being able to edit a short document repeatedly without retyping it in its entirety. Nevertheless, it would still be prudent to bear in mind the relative *cost* of preparing laser prints—which continues to be remarkably high despite the dramatic decrease in the price of the printers themselves.[13]

7.5 Laser Typesetters (Imagesetters)

The average reader would probably regard output from a standard laser printer as being of "high quality", but professionals are rarely satisfied with a resolution of 300 dpi, even taking into account improvements that might be achieved through photoreduction of laser copy prior to final printing (cf. Sec. 2.13). Close inspection of output from a laser printer clearly reveals "jagged edges" on the letters and on many of the lines used to create illustrations. Moreover, half-tone screens (Sec. 8.3) for a 300-dpi laser printer are crude at best, and uneven distribution of the toner often leads to inconsistent blacks.

It is for reasons like these that there has long been a desire to find a way to use electronic manuscripts as input for a *phototypesetting device* despite the complications noted in Section 1.1.5. Establishing the desired compatibility once required the availability of complex, unreliable, user-specific "translation software". A major breakthrough in resolving the interface problem came with the joint development by Linotype and Apple of a raster image processing (RIP) device to serve as a connecting link between a MACINTOSH computer and a Linotype phototypesetter. The RIP was designed to accept PostScript-language instructions issued by the computer and trans-

---

[13] The cost of printing a single letter-size page with a typical laser printer has been estimated at about $0.09–0.12. This assumes a purchase price of $1500 amortized over 3 years, 1000 printed pages/month, a toner-cartridge lifetime of 4000 pages (at a cost of $85), and typical outlays for paper and electricity. Significant savings are possible if toner cartridges are recycled rather than discarded, but it is important that one establish the reliability of a particular recycling service before entering into any long-term commitment.

form them page by page into commands recognizable by the typesetting device (also known as an *imagesetter*). This venture was so successful that now virtually any electronic document that can be printed with a LASER-WRITER can also be reproduced with a resolution several times greater directly on photographic film or photosensitive paper by the imagesetter's helium–neon laser (cf. also Fig. 6–2).[14] As a result, the way was effectively opened to applying desktop publishing techniques to the mass production of documents demanding the highest possible print quality.[15]

The simplest physical approach to preparing high-resolution copy consists of a MACINTOSH computer coupled directly to an imagesetter by way of an RIP (cf. Fig. 7–5).[16] The RIP intercepts the relevant PostScript commands as they emerge from the computer, translates them one page at a time into bitmap form at the appropriate level of resolution, and passes the re-

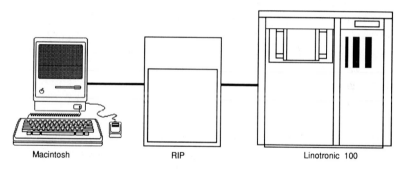

Macintosh   RIP   Linotronic 100

Fig. 7–5. Physical arrangement of the components required for obtaining imagesetter output of an electronic manuscript.

---

[14] The early LINOTRONIC Model 200 offered a resolution of 1690 dpi, but resolutions as high as 2540 dpi soon became available with the LINOTRONIC Model 300 (maximum page width 305 mm). The cost of the process is still in the range of several dollars per page, but this is not regarded as exorbitant given the large number of copies that can be prepared from a single LINOTRONIC "master". Unlike a laser printer, a phototypesetter has the ability to "print" to the very edge of a "page".
[15] Unfortunately, higher resolution is achievable only at the expense of throughput rate. Intervention of the RIP increases processing time by a factor of two to four relative to a LASERWRITER. The consequences are most apparent with text containing a wide variety of unusual fonts or complex PostScript graphics.
[16] Like laser printers, LINOTRONIC devices can also be incorporated into APPLETALK networks, extending the service to several clients simultaneously.

sulting information along to the imagesetter. The RIP is usually equipped with its own hard disk, a portion of which is reserved for descriptions of the most frequently employed fonts while the rest remains free for the temporary management of bitmaps.

Use of an imagesetter should be seriously considered in any situation requiring:

- output of the highest possible quality
- a rather large print run
- copy that contains text set in unusually small type or text displayed against a gray background
- large or odd-sized pages

Important points to bear in mind include the following:

- Imagesetter output should never be requested before corresponding output *(proofs)* from a laser printer has been subjected to meticulous inspection.

- Routine proofreading and copy correction should always be conducted in advance, at the laser-printer stage.

- Final laser-printer proofs should be readily available for consultation during the imagesetting process (cf. the comments regarding WYSIWYG output in Sec. 6.3).

- Optimum gray-screen settings for an imagesetter are different from those for a laser printer. A 10% screen (cf. Sec. 8.3) produces a surprisingly dark text background with a laser printer, but a value of 20% or more is more likely to be required with an imagesetter. Imagesetter grays appear lighter as a result of the considerably smaller dots used to construct the corresponding screens.

- Imagesetter output of graphics can be a very time-consuming process, especially non-PostScript graphics. The result may be a several-fold increase in page charges.[17]

- Not all documents benefit to the same extent from increased resolution. Very acceptable results can be obtained with a document composed ex-

---

[17] For example, preprocessing the illustration in Fig. 11–9 with a LINOTRONIC 200 (resolution 1693 dpi) took more than 25 min, followed by a 10-min reproduction step.

To expedite the printing of your disk, please fill in all pertinent information:

Disk Printing Order

Date: _____
Time: _____ Are we printing this job? ❏ Yes ❏ No

Job Title _____ ❏ New Publication ❏ Revision ❏ Standing Order

General Info:

NOTE: Disks are billed on a per-page basis and are usually run within 24 hours. Files are printed with a 10 minute time limit per page with additional preparation and/or printing time billed at current typesetting rate. Incomplete information on disk order form may result in additional charges.

RUSH JOB SERVICE FEES:
1-2 hours 100%
2-4 hours 75%
4-8 hours 50%

Disk Info:

All information listed by file name. Specify the kind of output you would like. List all fonts used in text and/or graphics.

Note: Please include TIFF, PICT, or EPS files on disk to ensure printing of graphics.

For PrePrint users:
1. Be sure colors in PageMaker match those in Illustrator and/or Freehand.
2. Colors knock out unless an overprint is specified.

(814) 865-7544

Business Services
Printing Services
The Pennsylvania State University
101 Business Services Bldg.
University Park, PA 16802-1002

PENNSTATE

Budget #	Our job #	U.Ed. #	Store Job ❏ Yes ❏ No
Department	Contact	Designer
Address		Phone
Billing Contact	Return Disk to:	
Billing Address		Phone

File #1 File Name _____ Software Application (specify version): _____

Page Size ❏ Tall ❏ Wide   Pages to Output   Quantity of each

Output to: **Fonts:**   **Color Separations:**

❏ Linotronic (1270 dpi unless noted below) _____   ❏ PageMaker spot color (files without color separated graphics)
  ❏ Paper ❏ Film negative   ❏ Knockouts in PageMaker
  ❏ 2540 dpi (for scanned photos only)
❏ QMS Colorscript Printer   ❏ Aldus PrePrint (Pagemaker with imported color graphics)
  ❏ Paper ❏ Transparency   **Colors:** ❏ Four-color separations   ❏ Adobe Separator (for Illustrator files only) Version: _____
❏ Laserwriter   ❏ Spot colors   ❏ Freehand Separations
                                List colors: ❏ Quark XPress Separations

Special Instructions: _____

File #2 File Name _____ Software Application (specify version): _____

Page Size ❏ Tall ❏ Wide   Pages to Output   Quantity of each

Output to: **Fonts:**   **Color Separations:**

❏ Linotronic (1270 dpi unless noted below) _____   ❏ PageMaker spot color (files without color separated graphics)
  ❏ Paper ❏ Film negative   ❏ Knockouts in PageMaker
  ❏ 2540 dpi (for scanned photos only)
❏ QMS Colorscript Printer   ❏ Aldus PrePrint (Pagemaker with imported color graphics)
  ❏ Paper ❏ Transparency   **Colors:** ❏ Four-color separations   ❏ Adobe Separator (for Illustrator files only) Version: _____
❏ Laserwriter   ❏ Spot colors   ❏ Freehand Separations
                                List colors: ❏ Quark XPress Separations

Special Instructions: _____

File #3 File Name _____ Software Application (specify version): _____

Page Size ❏ Tall ❏ Wide   Pages to Output   Quantity of each

Output to: **Fonts:**   **Color Separations:**

❏ Linotronic (1270 dpi unless noted below) _____   ❏ PageMaker spot color (files without color separated graphics)
  ❏ Paper ❏ Film negative   ❏ Knockouts in PageMaker
  ❏ 2540 dpi (for scanned photos only)
❏ QMS Colorscript Printer   ❏ Aldus PrePrint (Pagemaker with imported color graphics)
  ❏ Paper ❏ Transparency   **Colors:** ❏ Four-color separations   ❏ Adobe Separator (for Illustrator files only) Version: _____
❏ Laserwriter   ❏ Spot colors   ❏ Freehand Separations
                                List colors: ❏ Quark XPress Separations

Special Instructions: _____

Fig. 7–6. A typical request form for commercial output services.

clusively of text at a resolution of 1000–1700 dpi. On the other hand, copy containing half-tone images or "magazine-quality" four-color separations should be reproduced at a resolution of at least 2540 dpi, achievable only with the best imagesetters.

• The cost of imagesetting services varies widely, and is usually established on the basis of a detailed *estimate*. Obtaining an estimate requires that one be in a position to specify clearly the precise nature of the proposed job, including such information as a list of the required fonts, the file formats and sizes of any embedded graphics, the desired level of resolution, and the type of output desired (film or photographic paper; positive or negative; standard or mirror-image). A typical form for requesting commercial output services is illustrated in Fig. 7–6.

7.6 PostScript

7.6.1 Introduction

Despite its widespread acceptance, the PostScript language is actually of quite recent origin: PostScript was developed by programmers at Adobe Systems Incorporated (founded in 1982) and made its formal debut in 1985 along with the first LASERWRITER. The sole purpose of PostScript is to permit concise but unambiguous description of a printed page, facilitating direct transfer from a computer to a suitably equipped printer of all the information required to generate a full page of carefully formatted text and graphics. PostScript is essentially a programming language, and like most such languages it is based on a limited set of commands derived ultimately from English. Commands are augmented as necessary with appropriate numerical parameters and then arranged sequentially in such a way that a special PostScript *interpreter* can transform the information into instructions intelligible to the output device (cf. the description of a hardware-RIP in Sec. 7.5). The Apple LASERWRITER was one of the first commercial printers designed from the outset to accept PostScript instructions.[18]

---

[18] As noted in Sec. 6.3, attempts have also been made to develop an analogous *display PostScript* as a way of improving the fidelity of images presented on a

The PostScript language is *device-neutral*: its syntax and command set are entirely independent of the nature of the computer used to issue a set of PostScript instructions, and equally unaffected by the structure and resolution of the output device. A given PostScript description will produce identical results with any PostScript-compatible laser printer, ink-jet printer, or imagesetter (apart from qualitative differences that necessarily accompany differences in resolution). A PostScript program normally is "run" by a microprocessor in the output device itself rather than in the host computer.

A relatively short PostScript program can be capable of defining a surprisingly large and complex graphic image. Indeed, one of the greatest advantages of PostScript is that it requires a minimal amount of symbolic information transfer between the computer and the output device. This characteristic, together with the "built-in intelligence" of PostScript printers, explains why PostScript is held in especially high regard by network supervisors.

Every printer manufacturer now has the legal right to market devices equipped with PostScript interpreters. The language itself resolves problems that would otherwise inhibit communication between a computer and a printer, providing instructions that can be interpreted unambiguously by any suitably equipped output device. The PostScript page-description language has developed into something of an industry-wide standard for graphics-oriented printing, and it is strongly supported by most of the leading equipment manufacturers, including those that produce high-resolution image-setters (cf. Sec. 7.5). Software houses like Microsoft, Aldus, and Lotus have also enthusiastically embraced PostScript as a valuable medium for data interchange between various graphics applications.

Like the MACINTOSH computer itself, PostScript recognizes no fundamental distinction between letters of the alphabet and graphic images; text is handled formally as if it consisted of nothing more than a set of pictures. Type fonts are managed not as collections of bitmap images, however, but rather as mathematical descriptions of geometric forms ("outlines"). All the component letters, numbers, and symbols thereby become potentially avail-

---

monitor. The first computer that actually utilized a display PostScript was the ill-starred NEXT system, conceived by Steve Jobs, one of the founders of Apple. ADOBE ILLUSTRATOR is one of a small number of application programs that regularly uses PostScript to create some of its screen displays as a way of increasing the correspondence between what one sees and what later will be printed.

able in any desired size and fill texture on the basis of a few user-supplied parameters. This is one of the reasons why problems seldom arise with PostScript files containing a combination of text and graphics.

Certain PostScript-compatible graphics programs (e.g., CRICKETDRAW; cf. Sec. 11.4) offer the ability not only to send PostScript output to a printer but also to generate printed lists of the corresponding PostScript commands (see, for example, Fig. 11–10), a useful feature for anyone interested in learning more about the PostScript language by studying custom-tailored examples.

7.6.2 Simple PostScript Commands

PostScript is the most familiar example of a computer *page-description language* (PDL; cf. also Sec. 6.6). Its structure closely resembles that of other programming languages, but PostScript is unique in providing a rich stock of special operators for the manipulation and placement within a predefined space of both text elements (with any desired size and orientation, and in every conceivable font) and graphic objects constructed from any desired combination of points, lines, and curves.[19]

PostScript can be described as a "high-level" programming language, since it supports the inclusion of such complex structures as program loops and conditional statements. Like most other programming languages it recognizes *variables*, several types of *data*, complex *procedures*, and *control structures* of the form

```
if ... then
```

and

```
while ...
```

The language operates in a "postfix" or "reverse Polish" mode, which explains the first syllable of its name. This somewhat counter-intuitive approach

---

[19] A brief introduction to the PostScript approach to font definition is provided by Grosvenor, Morrison, and Pim (1992). More detailed information regarding the PostScript language generally is available from the books by Ross Smith (1990), Braswell (1989), Holzgang (1987), and Vollenweider (1988), as well as in various publications from Adobe Systems (e.g., 1986, 1988), especially the second edition of the *PostScript Language Reference Manual* (1991).

to expressing an arithmetic operation will already be familiar to many readers accustomed to working with certain types of pocket calculators (e.g., programmable calculators from Hewlett-Packard) as well as those acquainted with the programming language FORTH. A postfix language is characterized by operators placed not *between*, but *after* the corresponding operands. In other words, instead of the usual ("infix") sequence

$$1 + 2$$

a PostScript statement takes the form

```
1  2  add
```

where add is an *operator* specifying addition. The PostScript language recognizes more than 200 such operators, a few simple examples of which are presented in Table 7–1.

Table 7–1. Representative examples of operators recognized by PostScript.

Operator	Result	Operation
a neg	$a \cdot (-1)$	Inversion of the sign of a
a abs	$|a|$	Absolute value of a
a sqrt	$a^{1/2}$	Square root of a
a b add	$a + b$	Addition
a b sub	$a - b$	Subtraction
a b mul	$a \cdot b$	Multiplication
a b div	$a \div b$	Division

The postfix command structure operates on data stored previously in what is called a "stack". All data required for a particular operation are first introduced in proper sequence (from the "bottom") into the appropriate stack, after which the requisite operator is called. This operator then "withdraws" the required amount of information from the stack one piece at a time, acting upon each item as necessary to achieve the desired effect. The result is subsequently moved back into the stack, where it replaces data from which it was derived.

Postfix languages have the advantage that they are subject to very rapid interpretation by most computers, many of whose internal operations are also conducted in "postfix mode". An additional advantage is the fact that postfix program statements seldom require parentheses—a common source of programming errors. Thus, ambiguity never exists regarding the order in which a particular set of operators is to be applied: PostScript commands are always executed in the sequence specified, proceeding from left to right. Consider, for example, an expression of the form

$$2 + 3 \times 4$$

According to the standard rules of arithmetic (which confer a priority status on multiplication) the correct result of this set of operations would be 14, whereas straightforward *sequential* execution [i.e., $(2 + 3) \times 4$] would lead to a very different value: 20. Two equally valid ways of achieving the correct result (14) in PostScript are:

```
3 4 mul 2 add
```

and

```
2 3 4 mul add
```

An entirely different statement would be required to generate the value 20, such as:

```
2 3 add 4 mul
```

One characteristic that PostScript shares with FORTH is a straightforward approach to the definition of a new operator. Thus, if a particular application would be facilitated by an operator "cm" for converting centimeters to points, such an operator could easily be created with the simple statement

```
/cm{28.3465 mul} def
```

The *name* of the operator subject to definition (in this case "cm") is always preceded by a slash. The command "def" tells the PostScript interpreter that the operator specified is henceforth to be understood in terms of the expression in braces. Once such a statement has appeared as part of a program, a subsequent statement like

```
12 cm
```

will automatically be interpreted as meaning

```
12 28.3465 mul
```

(i.e., $12 \times 28.3465 = 340.158$).

A *page* (also known as the "default user space") within the context of PostScript is understood to be a planar Cartesian coordinate system with its origin (0,0) located at the lower left corner. The *x*-axis therefore coincides with the bottom edge of the page, while the *y*-axis represents the left side. Any particular point on such a page can be specified in terms of an appropriate *x,y*-coordinate pair, where both *x* and *y* are positive. The standard unit of measure in PostScript is the point (1/72″; cf. Sec. 2.4), so a line described as having the length "72" would be precisely one inch long.

7.6.3 Graphics Commands

Among the most important commands in PostScript are those that alter the dimensions and origin of the user space (the currently active coordinate system), permitting reduction or enlargement of a graphic object to be readily accomplished by stretching or shrinking the corresponding coordinate axes.[20] Three PostScript operators,

> translate, scale, and rotate

serve respectively to displace the origin of the coordinate system, alter its dimensions, and reorient the axes. Examples of effects that can be achieved with these basic operators are provided in Fig. 7–7.

Two additional operators,

> gsave and grestore

make it possible to create a record of the current graphic state of the overall system, or reset the system to its original state. These are often used as "clamps" to isolate some particular portion of a PostScript program so that conditions can be changed temporarily, with subsequent restoration later of the original set of conditions.

Only a few of the graphics operators in a typical PostScript program actually lead to marks on the printed page; most are instead involved in the construction of *paths*, or boundaries. A particular path is a discrete set of lines, geometric forms, and Bézier curves (cf. Sec. 11.4) that simply *describes* an object without actually reproducing it. A given path may or may not be

---

[20] The same strategy is applicable in PostScript-oriented graphics programs, such as ILLUSTRATOR or CRICKETDRAW (cf. Sec. 11.4).

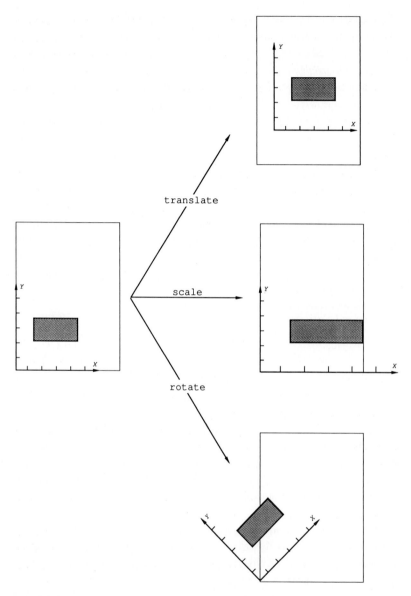

Fig. 7–7. Effects achievable with three of the basic PostScript graphic operators.

continuous, and it may or may not be *closed*. Some paths represent standard geometric forms (squares, perhaps, or ellipses), but others might specify sets of non-contiguous elements, or simply straight or curved lines, each with its distinct endpoints.

A PostScript program usually consists of several independent program segments, with each segment devoted to its own path. At the end of a given segment the current path would normally be *stroked* (printed)—leading either to a set of lines (possibly in the form of an outline) or to a surface "filled" with some type of pattern. The next program segment, containing a new path, would then be addressed. The term "current path" always refers to a particular set of points, lines, and curves that is *active*, and that describes a particular form assigned to a specific location on the current page. The active path is subject to selective distortion in a number of ways, which offers a convenient approach to creating new objects with unique shapes and/or sizes.

PostScript recognizes a distinction between three basic types of graphic elements: *geometric figures*, *text*, and *pictures*.

● Geometric figures

Geometric figures are created in PostScript as required with the aid of special graphics operators for describing straight lines and curves. These make it possible to define almost any outline or filled surface imaginable, and in any desired size and color (or gray shade). Such figures are manipulated on the basis of the corresponding *vectorial representations*.

Facilities are available for constructing figures from such geometric basis elements as straight lines, rectangles, circles, ellipses, and curves of various types; even letters of the alphabet can be treated as geometric figures. Each figure is associated with its own unique path, and is subject to transformation into either a *line* (solid, dotted, etc.) or *outline* of any desired weight, or into a two-dimensional *region* characterized by some type of screen pattern, which might or might not be surrounded by an outline. Regions with different fill patterns (or colors) can also be caused to overlap, and in any sequence specified. Several of the PostScript commands used in achieving such effects are listed in Table 7–2.

Table 7–2. Examples of PostScript operators and the effects they produce.

Operator	Effect, explanation
charpath	Incorporates the outline of a letter or other character into the current path (perhaps as a prelude to subsequent filling)
closepath	Closes the current path by joining the initial and final points with a straight line
fill	Covers the surface of some previously defined closed path with a single homogeneous color
findfont	Locates the set of character descriptions corresponding to a particular named font
grestore	Resets all parameters to a previously saved state (see gsave)
gsave	Creates a record in memory of the current set of coordinate-system parameters
moveto	Directs attention to a new location, defined by a specified pair of coordinates (x,y)
newpath	Resets the program (establishing an "empty path")
rlineto	Defines a line joining two specified points
rotate	Rotates the current coordinate system counterclockwise through a specified angle; the origin and the unit of distance remain unaffected
scale	Establishes a new unit of distance along the x and y axes without changing the origin or orientation of the coordinate system
scalefont	Associates a particular size with characters in the active type font
setfont	Sets the stage for a path to be based on the currently specified font at the specified size
setgray	Establishes a gray-scale value to serve as the basis for a fill pattern (0 black, 1 white)
setlinewidth	Defines the width (in points) of a particular line
stroke	Leads to physical output (printer, screen display, imagesetter) of the currently active path
translate	Moves the origin of the coordinate system to a new location; both the orientation and the unit of distance remain unaffected
true	Specifies that a preceding text argument (set in parentheses) is to be expressed graphically in terms of the currently active font (with a predefined location, orientation, size, etc.)

Fig. 7–8. Examples of variations that can be achieved with PostScript text.

● Text

Like any other PostScript "object", a letter of the alphabet is regarded as completely flexible with respect to its placement, size, potential distortion, and even reflection (cf. Fig. 7–8). A single PostScript statement suffices to reproduce a particular letter in any designated typeface based on any desired combination of size, line weight, thickness, and tilt. The quality of the printed results will be consistently high, in sharp contrast to letters constructed from a matrix of dots, where a change in size would lead only to a corresponding change in the size of the constituent dots. For this reason a PostScript-compatible printer has the inherent ability to reproduce an unlimited number of typefaces at any desired size and resolution. A particular face can also serve as the basis for many different geometrically defined character styles (shadow, oblique, etc.). Moreover, alphanumeric character output is not limited to the standard black-on-white format: text can also appear white against a black or gray background, or character outlines might be filled with distinctive patterns.

● Pictures

A "picture" in PostScript is understood to be a bitmap pattern. This might in turn be derived from a scanned image (a digitalized version of a photograph, illustration, or drawing), but it could also have its origin in a graphics program of the "paint" type (cf. Sec. 11.2).

Among the more important special effects that can be achieved with the aid of PostScript instructions are the following:

● text constructed from characters set at any desired angle and in every conceivable size (including fractional sizes; e.g., 7.35 pt)
● shaded type symbols or alphanumeric characters filled with complex patterns
● objects surrounded by outlines of any desired weight
● areas of gray with any desired intensity (e.g., 20.8%)
● shading *gradients*, covering any part of the range from white to black

Programming with the PostScript language is not particularly difficult, and PostScript programs can be prepared with any standard text editor (e.g., MICROSOFT WORD) for subsequent transfer to a printer. Most applications of PostScript are less overt, however, in that the desired set of effects is achieved visually with the aid of a graphics or page-makeup applications program that attends invisibly to the problem of communicating the user's wishes through the medium of PostScript.[21]

7.6.4 Sample PostScript Programs

Not every PostScript "program" is long and complex. Consider, for example, the following:

```
/cm {28.3465 mul} def
newpath
3 cm 2 cm moveto
1 cm 0 cm rlineto
0 1 cm rlineto
```

---

[21] As noted previously (Sec. 7.6.1), some programs, like CRICKETDRAW, permit one to examine directly a printed version of PostScript instructions that have been developed by the program itself in response to the user's graphic commands.

```
-1 cm 0 rlineto
closepath
.5 setgray
fill
stroke
```

Analysis of this set of instructions shows that it calls for output of a single square, 10 cm on a side, filled with a 50% gray pattern but without any solid border. The lower-left corner of the square is to be located 3 cm above the bottom of the page and 2 cm in from the left edge (cf. Fig. 7–9).

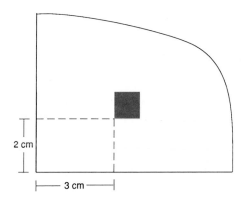

Fig. 7–9. Results obtained with the PostScript program described in the text for creating a shaded square at a particular location on a printed page.

Execution of the even shorter program

```
/cm {28.3465 mul} def
/Helvetica-Bold  findfont  120  scalefont
setfont
3 cm 4 cm moveto
45 rotate
(Draft) true charpath gsave .8 setgray fill
stroke
```

leads—as shown on the next page—to the word "Draft" in large (120 pt), boldface Helvetica type. The letters are filled with a 20% gray pattern,

Fig. 7–10. The "logo" that results from the set of PostScript commands listed in the text.

and the word is spread across the corresponding page at an ascending 45° angle.[22]

Finally, the complex "logo" shown as Fig. 7–10 was generated with the following short program, which includes the two custom operators "ShowOutline" and "WordCircle":

```
/Helvetica-bold findfont 12 scalefont
    setfont
/ShowOutline {true charpath stroke} def
/WordCircle {15 15 345
    {gsave rotate 0 0 moveto {PostScript}
ShowOutline
    grestore} for} def
216 216 translate
.5 setlinewidth
WordCircle 0 0 moveto (PostScript) true
    charpath
gsave 1 setgray fill grestore
stroke
```

---

[22] A program fragment like this one can be incorporated directly—and exactly as presented here—into any electronic document prepared with WORD (cf. Sec. 9.4). Assuming the instructions are expressly associated with the WORD paragraph style "PostScript", the corresponding graphic image will be reproduced as a *background* under one or more pages of text subsequently output by a PostScript-compatible printer.

8 Scanners

8.1 Introduction

Previously printed text and illustrations can both serve as valuable starting points in the course of preparing an electronic manuscript. Different approaches are required for carrying out the requisite transformations into machine-readable files, but each begins with the same piece of hardware: a *scanner*. The name "scanner" derives from the fact that the process by which most devices of this type capture information involves passing a beam of light in a linear fashion across the surface of the corresponding document.[1]

Our main focus in this chapter will be on scanners that interpret copy in terms of black, white, and various intermediate shades of gray. *Color scanners* have been relegated to a secondary status for the same reason that we have largely ignored color printers and color display screens: we are interested primarily in the preparation of black-and-white copy for scholarly publications in the natural sciences, where color is the exception rather than the rule. Moreover, DTP techniques for producing multicolored camera-ready (or color-separated) copy are still undergoing very rapid development. For the average user, color can fairly be described as a relatively expensive and complicated luxury, though the situation is likely to change dramatically in the years ahead.[2]

In order for copy to be scanned (text, a photograph, a drawing, etc.) it must first be subjected to light from a steady, concentrated source, usually

---

[1] Good general sources of information on the topic include the books by Busch (1991) and Glover (1989) as well as J. Martin's article "All About Scanners" (*Macworld*, October 1992, pp. 150–155).
[2] A harbinger of the future in this respect is the new bi-monthly magazine *Color Publishing* from the Advanced Technology Group.

some type of fluorescent lamp.[3] A large fraction of the incident light will be *reflected* from areas that are not intensely colored. The reflected rays are collected by an optical system consisting of mirrors and lenses, and then passed along to a "reading unit" made up of an array of photodiodes or semi-conducting sensors (CCD cells, "charge-coupled devices"). The surface of the original is probed point by point (or line by line) as a way of detecting regions of contrasting color, and the results are translated into electrical signals stored in the usual binary form. Scanning thus leads to a *bitmap representation* of the original document, where each point[4] on the scanned surface acquires a digital counterpart in the form of a stored bit of information (cf. also Sec. 8.2). In other words, a scanner translates a printed page into a digitized "mosaic". The higher the resolution of the scanner, the sharper and more detailed is the resulting bitmap.

Until relatively recently, scanners of two different types were available commercially:

- *graphics scanners*, designed exclusively for "reading" graphic material
- *text* or *OCR scanners* (OCR: Optical Character Recognition)

The mechanical requirements are essentially identical for the two technologies, so most scanners are now general-purpose devices well-suited to both applications—assuming access to the appropriate software (see below).

Once a document has been transformed into a digital representation at a specified level of *resolution* (cf. Sec. 6.2), the work of the scanning device itself is complete. What remains is the challenge of *processing* the captured image, an assignment best left to the host computer. In the case of *text scanning* this entails close scrutiny and analysis of the digital image in an attempt to recognize sets of points that might be interpreted as letters or other meaningful characters. The characters themselves must then be *identified* and *reassembled* as a suitably formatted text file (cf. Sec. 8.5). The nature of the processing required for a *graphic* image is a function of the intended purpose as well as the extent to which characteristics of the original must be modified, as discussed in Sections 8.2–8.4.

---

[3] The "overhead" scanner (see below) represents an exception to this generalization, since it operates on the basis of ambient light.

[4] The actual dimensions associated with such a "point" (or "sample") are a function of the resolution of the scanning device (see below).

The higher the scanning resolution, the greater will be the number of data points, and the more detailed the acquired information. Achieving a high degree of resolution presupposes use of a computer with a considerable amount of free memory for processing the accumulated data. Large images impose higher memory (and/or storage) demands than small ones: resolution at a level of 300 dpi (equivalent to standard laser-printer output) translates into ca. 14 000 data points per square centimeter of surface, so it is no wonder that memory constraints often constitute a severe bottleneck. A single letter-sized sheet analyzed at the 300-dpi level generates roughly 8×10^6 data points. Eight bits of black-and-white information can be recorded in one byte of memory, but a one-bit scan of a complete letter-sized page would still consume ca. 1 MByte of memory—clear evidence that a computer like the legendary "Fat Mac" with its "generous" 512 KByte of memory was ill-equipped for serious scanning applications.

A typical scanner today offers the choice of several scan resolutions, ranging from perhaps 75 dpi to 800 dpi or more.[5] Choices are also provided with respect to further processing and the subsequent storage of scanned images. DTP scans with a resolution of 300 dpi or more are usually saved under one of the two formats TIFF or EPS (cf. Sec. 8.4), both of which include provisions for retaining low-resolution images for display purposes along with the high-resolution data appropriate to the more exacting demands of a laser printer or imagesetter.

One of the most economical approaches to scanning takes advantage of a so-called *hand scanner*, such as the LIGHTNINGSCAN from ThunderWare or Logitech's SCANMAN. Hand scanners are ideal for anyone whose scanning requirements are infrequent and relatively modest. The active scan path with such a device is usually limited to a strip about 10 cm wide, but this is quite sufficient for many purposes, and the results can be surprisingly effective. Larger images are captured by using special software to "stitch together" multiple parallel scans.[6] A scan file is created simply by rolling the hand-held device (containing both a light source and a set of sensors) slowly and

---

[5] Scanning resolution is more accurately reported as "*samples* per inch" (spi), since "dots" are not really involved. Many scanners achieve high *apparent* resolutions (e.g., 1200×1200 "dpi") by *interpolating* between samples, a technique that may or may not lead to a significant improvement in image quality.

[6] The THUNDERWORKS program, supplied with the LIGHTNINGSCAN scanner (ThunderWare), offers exceptionally convenient and powerful tools for joining overlapping scans.

reasonably uniformly across a flat surface containing the image of interest (e.g., a photograph, a business card, a page within a book, or a clipping from a magazine or newspaper).

An even more economical—and quite ingenious!—early solution to the scanning problem was ThunderWare's pioneering product THUNDERSCAN, a system based on a tiny scan head that could be mounted easily in an Apple IMAGEWRITER printer in such a way that the printer was transformed temporarily into a *sheetfed scanner*.[7] Scanning with THUNDERSCAN was quite time-consuming, however, and the technique was applicable only to single-sheet documents that could be passed safely between the IMAGEWRITER's feed rollers. The low price of the unit (especially considering its 300 dpi resolution) was a consequence of the fact that the optical sensor was "piggy-backed" onto an existing drive mechanism and able to take advantage of communication lines established originally for an entirely different purpose. The THUNDERSCAN system might still be worth considering under certain circumstances—if one's needs were very limited, for example, and there happened to be a spare IMAGEWRITER available.

Most of the scanners marketed today are of the *flatbed* type, where the copy to be scanned is placed face down on the surface of a glass plate, just as with a photocopying device, permitting illumination and mechanical scanning from below. Less common is the *overhead* scanner, which in its design resembles a photographic enlarger. Here the sensor head is situated in a housing mounted several inches *above* the copy. One disadvantage of the latter arrangement is the difficulty of holding bound copy flat during the scan process. Very demanding scan applications may require use of a *drum scanner*, in which the copy is attached tightly to a drum programmed to rotate slowly past a precision scanning unit. *Slide scanners* represent a more recent development, permitting one to capture electronically images preserved originally in the form of transparencies.[8]

Scanners have recently been joined in the marketplace by an even more sophisticated technology for the capture of black-and white or full-color images: the *video camera*. Establishing communication between a video de-

---

[7] A THUNDERSCAN scan head occupies space ordinarily taken up by a print ribbon. A typical FAX machine might be regarded as another example of a sheetfed scanner, one specially adapted to telecommunication.

[8] See, for example, J. Matazzoni, "Seeing Through Slide Scanners" (*Macworld*, October 1992, pp. 156–163). Slide-scanning *attachments* are available as options with many standard scanning systems.

vice and a computer requires that the crude (analog) video signal first be *digitized*. One interesting product in this context is Canon's "still-video camera", a unit resembling a standard 35-mm photographic camera, but equipped to capture images on diskette rather than film.[9]

8.2 Scanning Graphic Images

As indicated above, a scanner translates any copy it encounters into an electronic bitmap at some predetermined level of resolution, which is then transmitted by cable to the computer. Assuming a scanned image is later to be reproduced with a 300-dpi laser printer, the corresponding scan should, in most cases, also be conducted at 300 dpi (see, however, the comments below on halftone screening). Figures 8–1 and 8–2 illustrate the kinds of results that can be anticipated when graphic images are scanned in this way and reproduced without any special processing.

In selecting a scanner for use in graphics applications it would be wise to conduct a series of scanning *tests* using one's *own computer and operating system*. Such tests should include at least one sample each of

- a line drawing,
- a continuous-tone image, and
- a halftone image,

as described below. The scanned images should be printed, saved to disk, and then subjected to some further manipulation within the context of appropriate page-layout and graphics software.

Virtually all the popular scanners today produce very satisfactory results with *line art*—graphic images composed exclusively of elements that are either black or white. Reproducing *continuous-tone* (black-and-white) images such as photographs is more complicated, and it requires a scanner capable of accurately distinguishing various shades of gray in order to assure

---

[9] Full-motion video presentations via computer can also now be realized by both MACINTOSH and WINDOWS users with sufficiently powerful computer systems. See, for example, Cary Lu's article "The Full-Motion Macintosh" (*Macworld*, January 1993, pp. 140–147).

a)

b)

c)

Fig. 8–1. Scanning electron micrograph of a polymer sample. a) Halftone reproduction of the original photograph; b) Image obtained when the original was scanned
as black-and-white line art at 300 dpi; c) Results from a 6-bit (64-level) gray-scale
scan of the same photograph (scan resolution 150 dpi), subsequently reproduced
with a 300-dpi laser printer using a 75-lpi screen.

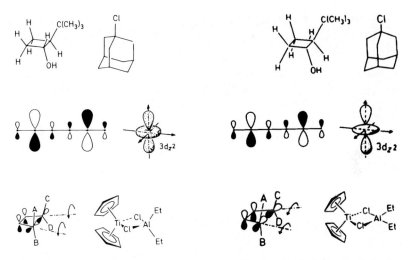

Fig. 8–2. Molecular formulas and text extracted from a book (line art). Left: reproduction of the original copy; Right: image obtained from a black-and-white scan at 300 dpi.

an image with adequate contrast. Achieving such *color resolution* means that every scanned "dot" must be tagged with an appropriate "grayness index" value. Prior to printing, the resulting information is then analyzed and transformed into a *halftone* representation (cf. Sec. 8.3), in which different shades of gray are simulated by more or less dense arrays of black dots. Problems often develop when one attempts to apply this technique to *reproductions* of photographs, or works of art that have been printed in books or magazines, because the resulting image is a "halftone of a halftone". This is the reason we suggest including (printed) halftone material in any test of a scanner's overall capabilities.

Tests of a scanner's *software* are just as important as those directed toward the device itself, since the best scanner in the world is useless without adequate tools for controlling it. For example, scan software should make it possible to specify precisely what portion of a document is to be scanned, and to establish an appropriate *threshold intensity* below which a scan signal will be interpreted as equivalent to "white". The controlling software should also permit one to adjust the resolution and save a scanned image according to the optimum format with respect to other important application programs (word processors, page-layout software, etc.). Among the most

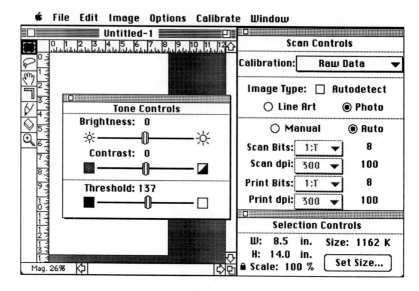

Fig. 8–3. The OFOTO user interface.

versatile file formats from this standpoint are PICT (or WMF), TIFF, and EPS, as discussed in Section 8.4.

Figure 8–3 illustrates the user interface and some of the controls available in the OFOTO scan program (from Light Source), features of which are described briefly below. The OFOTO interface appears again in Figure 8–4 in conjunction with the scan of a line drawing representing a molecular structure (enlarged with the aid of a "zoom" function).

An exciting recent development in scanner software is "one-button" control, implemented in conjunction with built-in intelligence that permits a scan system to "adjust itself" automatically for optimal results—somewhat analogous to working with an "automatic" 35-mm camera. Remarkably good digital images are generated in this way with little or no intervention on the part of the user. An example is the OFOTO software (bundled with the Apple ONESCANNER but also available separately from Light Source), which uses a rapid "pre-scan" to establish the nature and dimensions of the original, ascertaining at the same time whether or not the copy has been properly aligned on the scanning surface. Appropriate corrections are then applied as needed to assure the best possible set of graphic data, taking into account

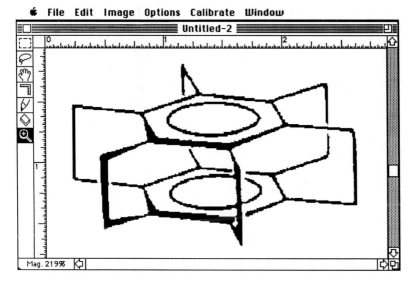

Fig. 8–4. Screen image (magnified) of a molecular formula scanned with OFOTO.

the specific printer that will be used for reproducing the image. (The user of course retains the freedom to adjust most of the parameters manually on an individualized basis.)

We conclude our introduction to graphics scanning with a few observations related to the use of *color* (including gray) in desktop publishing. A wide range of software is now available for preparing and editing documents containing polychromatic images, with tools that lend themselves to limited application even by users restricted to a monochrome monitor (though color and gray-scale monitors are becoming an increasingly common sight even in the historically black-and-white MACINTOSH world). Many of the scanners now on the market are able to capture quite faithfully even the most subtle color tones. An important consideration with respect to the subsequent reproduction of multicolor scans is the extent of *color depth* in the resulting file, as reflected in the number of bits available for color (or gray-scale) definition. The electronic images themselves consist as always of bitmap files, but it is possible for individual data points to be specially marked to reflect particular hues. A color depth of "4 bits" implies that every point is represented by 4 bits of information, with the potential for expressing $2^4 = 16$

different colors or tones. A continuous-tone black-and-white photograph can be reproduced as it might appear in a typical newspaper on the basis of roughly 6 bits of information per point (or *pixel*), but true "photographic quality", like that found in a book or magazine, requires 8-bit graphics permitting 256 (= 2^8) distinct shades of gray. Eight bits of information are also sufficient to produce multicolored images with color transitions more subtle than can be perceived by the average eye, but markedly better results are achievable with eight bits of information per point for each of three primary colors (e.g., red, green, and blue, the *RGB colors*). Treatment at this level opens the way in principle to 16.8 million (2^{24}) color possibilities for every dot.

Color scanners like the Hewlett-Packard SCANJET IIC, Apple's COLOR ONESCANNER, or the more economical Epson ES 600C achieve color depths of 3×8 bits through a scanning regimen equivalent to combining the results of three gray-scale scans. Filters or multiple sources of illumination are used to permit a single set of light sensors (often in the course of three successive scans[10]) to assign separate intensity values with respect to each of the three primary colors red, green, and blue. Not all software packages are able to support full-color processing at this sophisticated level, but such support is essential for preparing true *four-color (CMYK) separations*[11] of the type familiar to the printing industry.

8.3 Image Processing

Scanners and their associated software suffice for *capturing* a visual image and translating it into electronic form, but additional software is often required for *refining* the results. Günder (1988) characterizes a computer with such capabilities as an "electronic image-processing system". Among the more

---

[10] One-pass color scanners produce comparable results by alternating the flashes from three different light sources. For more information on color scanning see the article "Low-Cost Color Scanners" by C. Seiter (*Macworld*, November 1993, pp. 128–135).

[11] In "four-color separation" an RGB image is dissected into *four* basic print colors: cyan, magenta, yellow, and black ("CMYK"). Full-color pictures are reconstructed by placing tiny dots of inks based on these same four colors in such close proximity on the printed page that the eye perceives a spectrum of pure colors.

powerful MACINTOSH software packages for manipulating graphic images are DIGITALDARKROOM (Aldus), IMAGESTUDIO (Fractal Design), and PHOTOSHOP (Adobe), which unlike the others accommodates full-color images and is available for WINDOWS users as well. Other high-end WINDOWS packages include PICTURE PUBLISHER (Micrografx) and NUANCES (Fovea Expert). As the names imply, such programs effectively transport the user into an image-processing environment in which characteristics such as format, contrast, and brightness are all subject to individual adjustment, with special facilities for retouching, screening, and the application of a wide range of custom effects. Once a satisfactory result has been achieved the modified image is saved in a format that lends itself to printing or incorporation into a more complex document (cf. Chap. 10).

Figure 8–5 depicts the MACINTOSH user interface of Adobe's PHOTOSHOP. As with OFOTO (Fig. 8–3) and numerous other graphics applications, the usual menu bar has here been augmented by a "tool palette" offering a number of electronic "tools" for isolating and manipulating selected portions of a captured image. Also shown are a brush palette for adding graphic elements and a color palette for selecting or modifying color tones. Another option

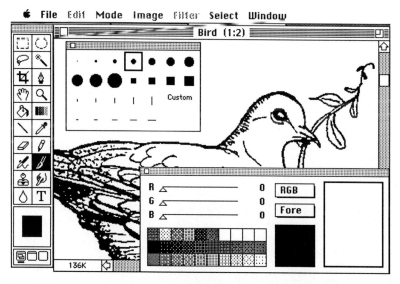

Fig. 8–5. The user interface of ADOBE PHOTOSHOP (version 2.0).

makes it possible to shrink an image or enlarge it by as much as a factor of 16, facilitating "microsurgery" at the pixel level.

At this point it is appropriate that we digress briefly and consider in more detail the way continuous-tone images are rendered by a laser printer—essentially a dot-matrix device limited to the two "colors" black and white. Gray shades can only be *simulated* under these circumstances. The usual approach is to redefine the image in terms of "super-pixels" made up of smaller dots, some of which are filled with black while others remain white. Thus a square 4 × 4 "super-pixel" can be used to simulate any of 15 shades of gray as well as black (cf. Fig. 8–6). Unfortunately, implementing this seemingly straightforward principle can exact a heavy price with respect to resolution. For example, a 300-dpi laser printer is capable of producing only 75 such "super-dots" per inch, so its effective resolution is reduced roughly to the level of a MACINTOSH screen display (or an IMAGEWRITER operating in "normal" mode). Moreover, a photographic image defined in terms of only 15 shades of gray would still be rather crude regardless of its resolution ("newspaper quality"). More gray shades can be accommodated by increasing the size of the "super-dots", but only at the expense of a further decrease in resolution. The usual compromise is either the one in our example, leading to images based on a "75-lpi (*lines*-per-inch) screen", or a 53-lpi screen with approximately 32 gray shades.[12] It would be pointless to scan a con-

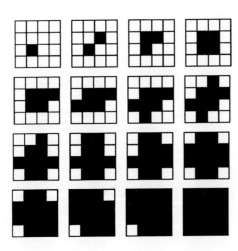

Fig. 8–6. One of the many possible sets of dot-blackening patterns (based on a 4 × 4 matrix) used for expressing gray shades by the halftone technique (adapted from Busch, 1991).

tinuous-tone image at very high resolution, since doing so leads to an enormous file from which most of the data will in any case later be discarded.[13] Figure 8–7a shows an 8-fold display-screen enlargement from an 8-bit, 150-dpi scan of a photograph in which the different color levels of the (scan) data points have been simulated by patterns. In Fig. 8–7b the same data are illustrated after transformation on the basis of a 75-lpi (circular) halftone screen.

The subject of screening is actually a rather complicated one, since different screen *patterns* (circular, elliptical, square, etc.) provide different effects, and screens can be applied to images in many different ways. An alternative approach to simulating gray shades is *dithering*, in which dots are applied more or less randomly, but with an appropriate *density*, rather than in the form of coarse "super-pixels. The difference is illustrated by the examples in Fig. 8–8.

The considerations above are also applicable to the problem of printing uniform regions of gray—as backgrounds, for example, or for distinguishing particular regions in a schematic diagram. The *gray value* that results from application of a particular screen pattern is usually expressed as a percentage, where "100%" represents black and "0%" corresponds to white. Specification of a "20% screen" might seem to be sufficient for communicating a desired level of grayness, but the *coarseness* of the screen (in lpi) and the *nature* of the individual screen points are important considerations as well, since it is a subtle interplay among the three factors that is ultimately responsible for the overall visual effect. Selecting the optimal screen for a particular situation often becomes a matter of trial-and-error, especially for the novice.

---

[12] Most of the world's professional printers classify screens not on the basis of lines/inch, but lines/*cm* even though DTP software (usually of American origin) provides screens rated in lines/*inch*. If a European publisher were to request use of a "30-(line) screen" the message should be interpreted for DTP purposes as requiring a screen setting of 75 (roughly equivalent to 30×2.54).

[13] This advice obviously does not apply to an image that will be reproduced with a high-resolution imagesetter (cf. Sec. 7.5), a device capable of producing much more satisfactory halftones. The best results are usually achieved with a scan resolution approximately twice the screen frequency (e.g., a 150-dpi scan for a 75-lpi screen). For more information on halftone scanning see the two excellent articles by S. Roth "Halftones Demystified" (*Macworld*, February 1993, pp. 175–180) and "Make Great Scans" (*Macworld*, February 1994, pp. 142–146), as well as the books by Blatner and Roth (1993) and Gosney, Dayton, and Chang (1991).

a)

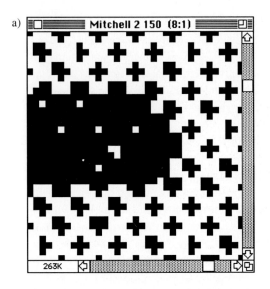

b)

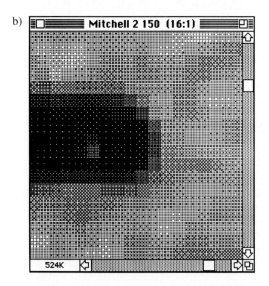

Fig. 8–7. Crude 8-bit scan data (a) in which different gray levels have been represented by geometric patterns, together with the result of transforming the information (b) into halftone form (for details see text).

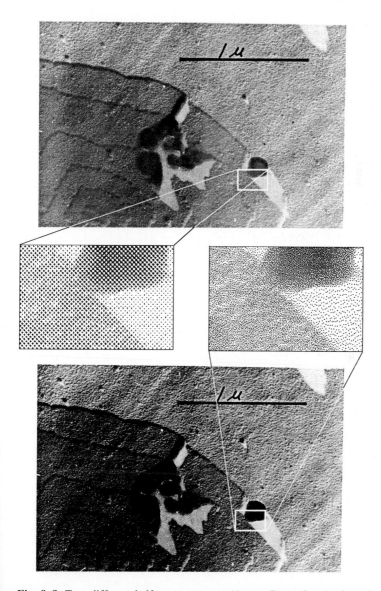

Fig. 8–8. Two different halftone treatments (ADOBE PHOTOSHOP) of an electron micrograph of a crystalline polymer in the process of dissolving in a solvent. Both images are based on the same 150-dpi, 8-bit scan, and both were printed with a 300-dpi laser printer. The halftone at the top was obtained with a 75-lpi, 45° circular screen, whereas that at the bottom resulted from diffusion dithering. (Original photograph kindly provided by D. J. Mitchell, Juniata College.)

8.4 *Graphic File Formats*

Selecting a particular *graphic format* establishes the way in which data corresponding to a digitized graphic image will be stored.[14] Most text and graphics programs operate internally on the basis of proprietary formats, ones that are not subject to interpretation by other programs. Some are quite unique, with no close analogies in other software, whereas others represent variations on common themes, perhaps modified by the application of special file-compression algorithms. Nevertheless, most programs provide the option of expressing graphic data in more or less standardized ways, facilitating the transfer of images from one programming environment to another. In choosing a piece of image-processing software it is important to ascertain which file formats it supports (for both import and export), since this is the key to interapplication flexibility. Graphic data can usually be transposed from one format into another without significant loss in quality or detail. Thus, both the TIFF and EPS formats (see below) are fundamentally capable of expressing equivalent amounts of basic information. On the other hand, saving a MACDRAW image as a PICT file, for example, could result in loss of important information related to color.[15] Unfortunately, the subsequent act of *importing* stored data into a new environment occasionally does lead to unexpected results due to undocumented program idiosyncrasies. Problems of this type are often resolved most easily by experimenting with alternative transfer strategies, involving other programs and/or different file formats.

Several more or less "standard" formats have evolved for graphics files in the MACINTOSH environment, and a similar trend is apparent for WINDOWS applications as well, though the IBM-compatible world was long plagued by the very lack of such standardization. As a result, graphic data from many

---

[14] Similar considerations are applicable to text files as well (cf. Sec. 9.6). Note that the *format* associated with a graphics *file* is something quite different from the *nature* of the graphic *image*. A great many file formats have been developed over the years—some of which actually overlap—but there are only two fundamental *types* of electronic images: bitmaps and vector graphics (cf. Chapter 11).

[15] One of the most versatile tools for converting files from one format to another is ADOBE PHOTOSHOP, available in both MACINTOSH and PC versions. For a good discussion of the more general problem of MAC–PC file conversion see the article "Bridging Two Worlds" by C. Seiter (*Macworld*, March 1992, pp. 152–159).

sources can now be imported into a wide variety of word-processing, page-layout, and presentation settings. Several of the most common formats for graphics applications generally—including scanned images—are described briefly below in alphabetical order (cf. also Chapter 11; formats for text-file transfers are discussed in Sec. 9.6):

● Bitmap (or *raster*) format

"Bitmap format" is a generic term applicable to any file based on a pure bitmap image (e.g., PAINT or PCX files; cf. also Sections 11.1 and 11.2). Different programs are designed to process bitmap images at different resolutions, as illustrated by MACPAINT (72 dpi), SUPERPAINT (300 dpi), and CANVAS, which can handle various resolutions up to 2540 dpi.

● CGM

The CGM (Computer Graphics Metafile) format has been recommended as a standard for vector graphics by the International Organization for Standardization (ISO) and the American National Standards Institute (ANSI), but it has not yet been adopted by the MACINTOSH community.

● DCS

DCS (Drawing Color Separation) is not really a format, but rather a technique developed by Quark for producing TIFF and PICT color (CMYK) separations that can be imported more readily into various program environments.

● DRAW

DRAW refers to a particular object-oriented file *type* (DRWG), developed in conjunction with the program MACDRAW. The term is sometimes (incorrectly) used as a way of referring to *any* file containing a vector-graphic image, whereas it in fact implies a file based specifically on the PICT format (see below).

● DXF

Autocad's "Drawing Interchange Format" (DXF) provides a way of expressing high-quality, object-oriented graphics in ASCII form.

● EPS (Encapsulated PostScript)

EPS is a format designed for images created with a PostScript graphics editor (cf. Sec. 11.4). Like a TIFF file (see below), a typical EPS file includes not only a PostScript description of an image but also a low-resolution (PICT or .TIF) "screen version" for approximating the image on a monitor or a non-PostScript printer. The EPS format is supported by a wide range of applications, including ILLUSTRATOR, CANVAS, CRICKETDRAW, PIXELPAINT, and LASERPAINT. One important drawback to EPS files is that they have a tendency to become quite large and unwieldy (roughly twice as large as comparable RIFF files; see below). Also, not all EPS files have identical characteristics, a problem most likely to be encountered in conjunction with DOS-based programs.

A vector graphic that has been transferred from one program into another in the form of an EPS file remains subject to alteration with respect to its dimensions, and certain other deformations may be possible as well, but changes can rarely be made in the basic structure of the image itself (despite the fact that the image may sometimes be caused to *look* different on the screen as a consequence of changes made in the low-resolution companion image).

● FOTO

Some scanners generate files of the type known as FOTO. These represent nothing more than simple bitmaps, where each pixel is expressed by a single bit of information. The FOTO format was developed originally for version 1.2 of PAGEMAKER, but it is not supported by later versions of the program. FOTO files should therefore be translated into TIFF format for import purposes, for example with the aid of IMAGESTUDIO.

● GEM

GEM refers to a bitmap format for the PC developed in conjunction with the Ventura "Graphics Enhancement Manager", and characterized by the file extension ".IMG".

● GIF

The Graphics Interchange Format (GIF) is a compressed bitmap format devised by CompuServe to facilitate on-line graphic data transfer, though it is also subject to import into programs like ADOBE PHOTOSHOP.

● HPGL

"Hewlett-Packard Graphics Language" (HPGL) is used primarily as a way of defining images for output to a plotter.

● PAINT

As its name suggests, the PAINT format (file type PNTG) originated with the pioneering MACINTOSH graphics program MACPAINT (Sec. 11.2). Graphic information in a PAINT file is recorded serially, pixel by pixel, just as with any other bitmap file. Each pixel is represented by a single bit of information, and an image is always defined at a resolution of 72 dpi (equivalent to that of the MACINTOSH display screen). A PAINT file can thus be regarded simply as a digitized mosaic of black and white dots.

Storing a scanned graphic as a PAINT file may lead to a significant loss of resolution, and all gray-scale information will disappear as well. MACPAINT files are further restricted by a rectangular boundary measuring 8″ × 10″ (ca. 20 cm × 25 cm) and oriented vertically; any information that would exceed these limits is discarded during storage.

● .PCX

PCX is the native (bitmap) format for PC programs in the PAINTBRUSH family (from Z Soft), and it has long been a familiar part of the IBM-compatible world.

● PICT

The PICT format (not an acronym) was developed in conjunction with the first version of the program MACDRAW (Sec. 11.3). Its primary function is to convey vector-graphic information, though bitmap data can also be accommodated—in any of the eight QUICKDRAW colors and at any desired resolution (i.e., the 72-dpi limit of a PAINT file does not apply). PICT is also the format most often utilized by a MACINTOSH computer when a graphic image is saved to the operating system "clipboard".

● PICT2

PICT2 represents an extension of the original PICT concept, one designed originally to take advantage of the first MACINTOSH II's ability to deal with a wide array of colors and multiple shades of gray. Analogous to PICT, a PICT2 file may contain a combination of bitmap (at any resolution) and vector-graphic information, but in this case the information can be expressed in as many as 16.8 million colors (note, however, that an *8-bit* variant of PICT2 is restricted to 256 colors). Programs such as ADOBE PHOTOSHOP, DIGITALDARKROOM, APPLESCAN, and STUDIO/8 or STUDIO/32 (Electronic

Arts) work smoothly with the PICT2 format, which facilitates the modification of such data files irrespective of their origin.

● PostScript

A PostScript file is a text file capable of generating a graphic image on any device equipped with a PostScript interpreter (cf. Sec. 7.6). Unlike EPS files, pure PostScript files include no provision for displaying the corresponding image on a monitor. PostScript files are subject to editing with almost any word processor.

● RIFF (Raster Image File Format)

Contrary to what one might assume, RIFF is not a variant on the format TIFF (see below), but rather a unique format developed for Fractal Design's programs IMAGESTUDIO and COLORSTUDIO (and recognized by QUARKXPRESS and DESIGNSTUDIO, among others). It offers the advantage of requiring only about half the storage space of a corresponding EPS or uncompressed TIFF document. RIFF files support up to 256 gray values per data point (equivalent to a color depth of 8 bits).

● TGA

TGA (or TARGA, from "Truevision Advanced Raster Graphics Adapter") is yet another bitmapped graphics format utilized by some MS-DOS software in conjunction with Truevision image-capture boards.

● TIFF, .TIF

"Tagged Image File Format" (TIFF) was devised jointly (in 1986) by Microsoft, Aldus, and the developers of various scanners, and it is now supported by most commercial scanning software. TIFF files are also accepted by a wide variety of applications programs (e.g., PAGEMAKER, READYSETGO, ILLUSTRATOR, etc.). There are actually several variants on the basic TIFF format, differing with respect to the maximum number of gray values recognized.[16] Analogous—but not identical[17]—files in the PC world carry the label ".TIF".

---

[16] TIFF files are also subject to various *compression algorithms*, useful as a way of economizing on storage space. This has led to expressions like "4-bit TIFF, compressed" or "TIFF Packbits". File compression is discussed briefly in Sec. 5.5.

TIFF was originally intended to provide a common standard for all gray-scale scanners, but it has since evolved in various directions (e.g., monochrome TIFF, gray-scale TIFF, color TIFF), and the results are not always fully compatible from one application to another.

TIFF files are capable of preserving bitmap information at any desired resolution, and with various levels of color definition, but no facilities are provided for object-oriented images. A TIFF file often contains two different bitmaps of a single image: one at high resolution, consuming a correspondingly large amount of storage space, and the other providing a cruder representation consistent with the resolution of a standard screen display. The latter requires considerably less space, and is much easier to manipulate. In many cases, importing a TIFF file into a host document results in formal inclusion of only the smaller of the two segments, together with a record of the *disk location* of the complete file. In order to take advantage of such a "file link" the computer must of course be able to locate the appropriate TIFF file when the time comes to express the full set of information with a high-resolution printer; otherwise only a MACPAINT-like image could be printed. In many cases, successful linkage requires that the TIFF file be consigned to the same folder (subdirectory) as the document to which it relates.

● ThunderScan (Gray-Scale Table Format)

The ThunderScan file format is unique to the THUNDERSCAN scan program (cf. Sec. 8.1). Each pixel is associated with 5 bits of information, for a total of 32 possible gray shades.

● Video format

This broad term covers any format used in capturing information with a camcorder or similar video device.

● WMF

Windows Metafile (WMF) is the WINDOWS equivalent of PICT, with provisions for both bitmap and vector graphics.

---

[17] Fortunately, the MAC–PC distinctions are less serious with TIFF files than in the case of EPS, and file conversion from one platform to the other is usually relatively straightforward.

8.5 *Optical Character Recognition*

The goal of *optical character recognition* (*OCR*) is to locate and isolate alphanumeric information that may be present in a scanned image, and then to interpret this information and save it as an electronic text file. The challenge is far more demanding than it may sound, since it requires equating the *picture* of a letter (which may be subject to considerable distortion) with the *concept* of the letter itself (i.e., with a corresponding ASCII-code value). Successfully implemented, OCR can open the way to applying to a *printed* document all the powerful techniques of electronic text editing—just as though the original had been entered through the keyboard.

Interpretation of a set of bitmap images as alphanumeric symbols generally occurs within the computer, not the scanner, and success in the venture depends almost entirely on the sophistication of the corresponding software. Two basic approaches have been taken to solving the fundamental problem of optical character recognition:[18]

● *Similarity tests* based on fixed sets of letter patterns (also known as the *matrix-matching* or *pattern-matching method*)

With this methodology anything suspected of being a symbol in the original image is closely compared with a set of standard representations of the various letters constituting the alphabet (and appropriate supplemental characters). An attempt is then made to establish a best possible *match* within some predetermined span of tolerance. A certain amount of latitude is always provided to reflect the subtle variations that inevitably differentiate one example of a letter from another.

● *Feature recognition* (or *feature analysis*)

In this case the program attempts to assign an *outline* or a group of *geometric features* to a perceived character, and then to correlate this information with a set of distinguishing characteristics. For example, a character interpreted as a small circle resting on what appeared to be a baseline might be equated with the letter "o". Such an analysis based on distinguishing char-

---

[18] For recent reviews of OCR software see J. Heid, "From Pixel to Page" (*Macworld*, July 1992, pp. 174–181) and C. Seiter, "OCR: The Recognition You Deserve" (*Macworld*, November 1993, pp. 92–98).

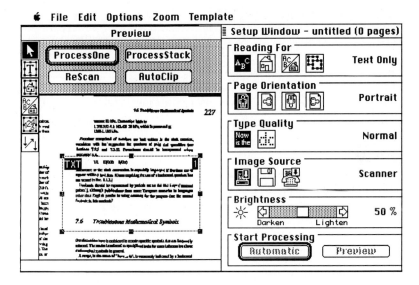

Fig. 8–9. User interface for the OCR program READIT! O.C.R. PRO 3.0A.

acteristics within a set of symbols would appear to represent the more fundamental and universal approach to character recognition—especially in conjunction with the principles of "fuzzy logic" (McNeill and Freiberger, 1993)—but trainable pattern-matching software has so far proven to offer the greatest promise for correct interpretation in the case of special symbols or text set in an unusual typeface.

One of the most widely used pattern-matching OCR programs for a personal computer is READIT! O.C.R PRO from Olduvai Software (cf. Fig. 8–9). The program package includes several standard "type tables" already optimized for the recognition of commonly encountered typefaces (typewriter script, a serif face like Times, type as it appears in several popular periodicals, etc.). A well-designed set of tools facilitates the creation of new type tables based on the user's own text samples. Thus, a simple menu command activates the program's "learn mode", in which the user is called upon to provide guidance in the interpretation of characters encountered in the first few lines of a piece of scanned text (Fig. 8–10). In this way the program gradually assembles a reliable new set of relevant bitmap patterns at the appropriate level of resolution, ultimately incorporating all the letters of the alphabet

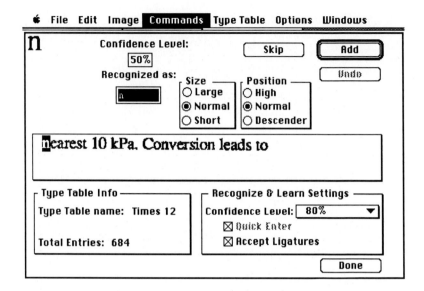

Fig. 8–10. An example of the way the "learn mode" in READIT! O.C.R. PRO is used for creating a type table. One particular perceived character has been highlighted in its scan context in the box at the center of the screen, and then reproduced in enlarged form at the top left corner. The proposed interpretation (subject to correction by the user) has been entered in the "Recognized as:" box, together with an indication of the certainty level applicable to this assignment. The user must now evaluate the interpretation and decide whether or not this example of the character should be incorporated into the program's type table for future reference.

(Fig. 8–11). By contrast, *omnifont* programs like WORDSCAN (Calera) and OMNIPAGE or OMNIPAGE PRO (Caere) begin immediately to interpret anything they see, though the OMNIPAGE PRO package does support optional training for unusual or difficult characters.

Several severe challenges stand in the way of routinely achieving satisfactory OCR results, especially given that scanned originals vary widely in quality and style:

● An effective program must have some way of accommodating significant variation in the *form* of a particular character, and provisions are required for dealing with alternative *type styles* (e.g., both roman and italic characters must be subject to recognition, either with or without serifs).

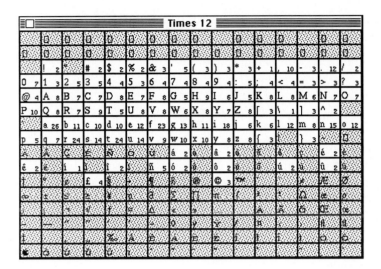

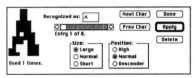 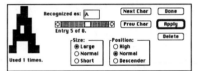

Fig. 8–11. A typical READIT! O.C.R. PRO type table, showing the number of specimens available for each character (where shaded characters are so far not represented). Below the table are two of the eight recorded specimens of the capital letter "A".

- *Kerning* and the presence of *ligatures* may make it difficult for a program to ascertain precisely where a given symbol begins and ends. "Broken" letters can also be a problem, particularly with typewritten or poorly printed documents; thus, a broken "m" might easily be interpreted as the combination "rn".

- Some fonts provide virtually no reliable clues for distinguishing between certain characters, such as the small letter "l", the numeral "1", and the capital letter "I".

- Some means should be available for dealing with *special characters* not part of the standard alphabet (mathematical symbols, arrows of various types, etc.), preferably involving user-defined equivalencies.

- Finally, a good general-purpose OCR program should be capable of capturing text set in *multiple columns*, or text containing *illustrations*, and provisions should exist for incorporating tabular data in *tab-delimited* form.

The probability of achieving a successful result with OCR depends heavily on the print quality of the scanned original. FAX documents are notoriously troublesome in this regard, and even under the best of circumstances characters with similar forms (e.g., "h" and "b", or "O" and "0") are frequently interchanged, so it is essential that an electronic document acquired in this way be proofread with great care. In working with pattern-matching software good judgment must be exercised in selecting the model character set—to ensure, for example, that the "training" document really does include a representative set of symbols. Progress during the learning process can usually be assessed by noting the frequency with which the program requests user intervention.

Small type and indices (subscripts or superscripts) are especially common sources of errors in OCR files; indeed, most of the programs currently available are incapable of assigning correct formatting to indices (a major disadvantage for anyone scanning text containing large numbers of chemical formulas!). For scientific applications it is especially important to select OCR software that at least supports the assignment of pseudonyms to special characters. Thus, if a program could be taught to *recognize* the Greek letter "Δ", for example, even if it were to interpret it as some other symbol (such as the rarely-encountered "@"), a subsequent search-and-replace routine (cf. Sec. 9.2) could be used to restore the correct symbol wherever it belonged.

The time necessary for scanning and interpreting a page of text (i.e., ca. 2500 characters) is heavily dependent on the quality of the original, varying from as little as 45 seconds to several minutes. An example of a scanned text image and the corresponding text as deciphered by WORDSCAN is provided in Fig. 8–12.

OCR software for personal computers has advanced to the point where it is now practical to capture printed text on a routine basis with an error rate below 1%—but this degree of latitude is still consistent with as many as 15–25 mistakes per manuscript page, and it presupposes relatively clean copy. The presence of numerous special symbols and equations may complicate the process to such an extent that it would actually be more efficient to recreate the target document manually. Reliable OCR software is also relatively

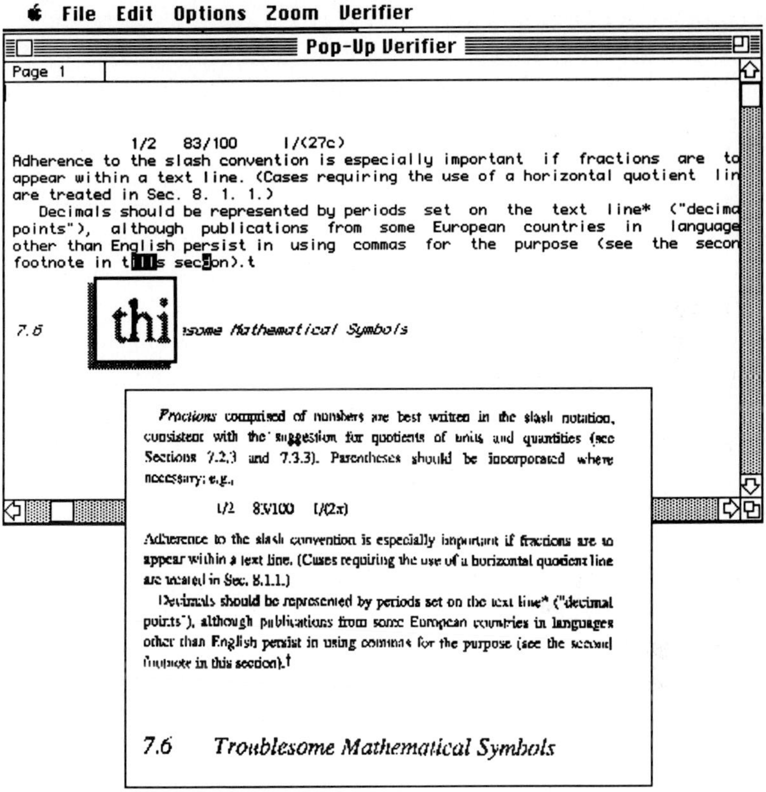

Fig. 8–12. Scanned text (inset), together with the way it has been interpreted by WORDSCAN PLUS. Note that "2p" in the first line has mistakenly been interpreted as "27c". The program has paused for confirmation at what it recognizes to be another trouble spot: where "this" has been interpreted provisionally as "tills".

expensive, and for all these reasons it is unlikely that the technique will prove cost-effective in the near future for the average user whose text-scanning needs are limited and infrequent.

Part III
Software

9 Word Processing

9.1 Introduction

The term <as used here is meant to include any computer program applicable to the solution of problems in the natural sciences.[1] In the broadest sense this would include programs for capturing experimental data, data simulation and optimization, database development, cataloging and otherwise managing literature references, processing graphic images (in either two or three dimensions), and—perhaps most important with respect to the daily activities of the average scientist—*word processing*, with special provisions for dealing with technical equations and formulas of various types.

The input phase of a word-processing operation is essentially analogous to the preparation of text with a convenient, full-featured typewriter—except that the newly created characters and symbols would normally not be expressed immediately on paper. The similarity ends here, however. Indeed, the most distinctive and useful characteristic of computer-assisted writing is the ease with which a developing document can be refined and adapted: chiseled and polished until it acquires its optimal and final form.

Even the most primitive of the early computers were equipped with *line editors* to assist programmers in the drudgery of correcting and revising their creative efforts. A line editor consists essentially of a set of formal commands for developing and modifying relatively brief strings of text, which can then be arranged to form organized *lists* for guiding the work of the computer itself. A modern *word processor* might be regarded as a dramatic evolution of this basic concept, where the goal has become that of providing the creative writer with powerful tools for preparing documents of all types, from brief

---

[1] A very comprehensive two-part review of scientific software appeared recently in two "Marktübersicht" sections of the German Chemical Society's news publication *Nachrichten aus Chemie, Technik und Laboratorium* **1992**, *40*, issues 6 and 9.

memoranda to full-length book manuscripts, complete with provisions for complex formatting and the inclusion of graphic material.

The software market features a bewildering array of word-processing programs, including several with cross-platform compatibility, and the novice could scarcely be expected to make an intelligent choice. Simply accepting the advice of a dealer is a risky strategy at best. In this and the next section we offer various criteria we believe should be borne in mind in comparing the characteristics of various programs, whereas Sec. 9.3 focusses on specific word-processor characteristics of particular relevance in the natural sciences.

It was noted earlier (in Sec. 1.3) that it is becoming increasingly difficult to draw a clear distinction between word-processing and *page-layout* software. Modern word processors claim many attributes that once were associated exclusively with layout programs, and the reverse is true as well. Irrespective of whether or not a piece of text will ultimately be transferred into a page-layout (page-makeup) environment, the following formatting features should today be regarded as mandatory and available for convenient use in the very first drafts of a manuscript:

- straightforward, easily accessible controls for the arrangement of printed text on a standard page (with respect to margins, line spacing, etc.)
- commands that make it easy to change typefaces and specify alternative type styles
- access to a wide range of special symbols, especially the letters of the Greek alphabet
- tools simplifying the preparation of complex tables
- facilities for integrating graphic images into a text document
- flexible import and export capabilities for both text and graphics
- pagination that is automatic but subject to some form of manual override
- clear screen indications of all page breaks

The following checklist[2] may prove useful in a preliminary assessment of the advantages and disadvantages of various competing programs.

1. Technical specifications

Minimum memory requirement	...KByte
Program fully resident in memory	y/n
Text displayed on the screen approximately as it will appear in print (WYSIWYG; cf. Sec. 6.3)	y/n
Copy protection	y/n

2. Constraints

Maximum file size (irrespective of storage medium)	...KByte
Maximum line length	...characters
Text (up to ...KByte) retained in memory	y/n
Compatibility with other programs	y/n

3. Screen display

Boldface, italics	y/n
Proportional fonts	y/n
Various sizes of type	y/n
Subscripts and superscripts	y/n
Realistic line spacing and margins	y/n

4. Special features

Mouse support	y/n
Commands addressable from both mouse and keyboard	y/n
Footnote and endnote management	y/n
Flexible headers and footers	y/n
End-of-line hyphenation (automatic or on demand)	y/n
Spelling checker	y/n
Indexing facilities	y/n
Form-letter function ("print merge")	y/n
Background printing	y/n
Arithmetic functions	y/n

---

[2] This list has been adapted from one that appeared in Special Issue 1/1987 of *Computer Persönlich*, which was devoted entirely to the subject of word processing. A similar approach was taken by S. Aker in her article "Key Words" in the "Buyer's Guide" section (pp. 18–28) of the June 1992 issue of *MacUser*.

9.2 Minimum Requirements for Convenient Text Preparation

Word-processing programs have matured dramatically in recent years. This section highlights several of the general characteristics and capabilities that can now be regarded as mandatory in a full-featured word processor.

- "User friendliness"

The most important consideration in selecting a word processor should be ease of use. Computer technology has advanced to such an extent that the benefits of electronic data processing can now be openly accessible to almost everyone—students, secretaries, busy executives, and of course working scientists. No one should feel compelled today to submit to a lengthy course of instruction in computer fundamentals as a prerequisite to preparing a straightforward electronic text document, and most of the steps involved should quickly become transparent and nearly intuitive.

At the same time, it would be unrealistic to expect a full-featured word processor to be both powerful and at the same time trivial to master. The ideal program should be accessible at a meaningful level after a very brief introduction—leaving the impression that one is actually working with a "clever typewriter"—but it should also be equipped with a wide range of far-reaching "extras", features that will only be assimilated gradually: as the need arises, and in conjunction with a careful reading of the user's manual.

- A well-written handbook

The availability of a good handbook or user's manual is essential for anyone aspiring to take full advantage of the potential inherent in any complex piece of software.[3] This manual should be designed to serve both as an ef-

---

[3] Despite the inherent truth of this observation we hasten to add that it is virtually impossible for one to master the complete contents of a good software manual immediately and in a systematic way, especially at the outset, when many of the features described will lie too far beyond one's perceived needs. It is always best when approaching a new program to start with the basics. With a word processor this means installing the software and then beginning immediately to type a simple piece of text, thereby acquiring familiarity with the basic menu structure and certain essential aspects of mouse control and data entry. Elementary steps related to routine

fective introduction to the program and as a comprehensive source of guidance for consultation later when specific questions arise. It would be difficult to overstate the importance of a clear *tutorial* section. The ideal tutorial is rather brief, and is structured around a series of well-conceived examples, each accompanied by a sufficient amount of detailed explanation. Working through a set of examples is the key to learning in the most effective way possible: by actually *doing*. Examples should also be provided to illustrate more abstruse and obscure features of the program, since these may otherwise prove frustratingly difficult to implement. The absence of a first-rate manual may not be a fatal shortcoming for a program—provided someone has published a sufficiently lucid book on the subject, as in the case of the Cobb Group Staff (1993) for the program WORD (Sec. 9.4).

● A reasonable "learning curve"

The most elementary technical aspects of writing—such as text entry, saving one's work to disk, and preparing backup copies—should be especially easy to accomplish, requiring little insight into the structure and scope of the program as a whole. Convenient provisions for saving a file are especially important. No word processor is perfect in this respect, but some programs make it less likely than others that a single careless move, occupying a fraction of a second, will negate several hours of painstaking effort.[4]

● Provision for multiple document windows

The *windows* approach to screen output (cf. Sec. 4.4) opens the way to simultaneous display of two or more segments of the same document (as shown in the case of WORD in Fig. 9–1), a feature that greatly simplifies the task of moving text from one location to another. Facilities may also exist for examining and manipulating the contents of two or more *different*

---

editing (adding, replacing, and deleting text, as well as saving and recalling a document) should be the next concern, at which point one is already prepared to undertake the first serious writing assignment.

[4] The latest version of WORD (Sec. 9.4), for example, includes automatic (optional) screen display of "save" reminders at predefined intervals. An interesting accessory program can also be of enormous help in this context: LAST RESORT from Working Software, which silently maintains a continuous record of every keystroke executed in the course of a particular working session. Retrieving a "lost" file on the basis of a LAST RESORT transcript can be somewhat time-consuming due to the inevitable presence of data related to corrections and other peripheral activities, but it is at least *possible*.

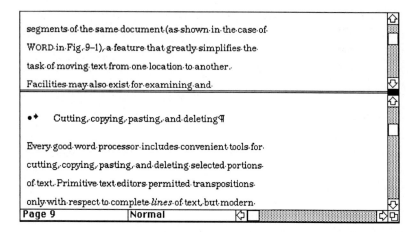

Fig. 9–1. "Split-screen" view of text prepared with MICROSOFT WORD, showing simultaneously two excerpts from the same document.

documents (cf. Fig. 9–2). In most cases only one of several "open" windows is *active* at any given moment (i.e., subject to immediate modification)—as indicated, for example, by "shaded" scroll bars—but an alternative open window can be activated instantly simply by moving the mouse cursor to a point within the boundaries of that window and clicking the mouse button. One of the many advantages associated with multiple windows is the ease with which a fragment of text can be copied from one source and then immediately "pasted" into another (see below).

● Cutting, copying, pasting, and deleting

Every good word processor includes convenient tools for cutting, copying, pasting, and deleting selected portions of text. Primitive text editors permitted transpositions only with respect to complete *lines* of text, but modern software lets the user operate on any set of contiguous characters, irrespective of length, with subsequent automatic adjustment of the lengths of all affected lines. It should also be possible to select with a simple mouse or keystroke operation discrete words, sentences, or paragraphs as a precursor to deletion, highlighting, or transfer of the corresponding element to a new location.

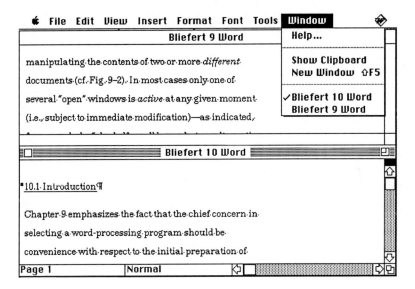

Fig. 9–2. A WORD display with two independent windows for working in two different documents. The shaded scroll bars indicate that the lower window is the one currently active.

● Search-and-replace routines

A "search" function enables the user to ascertain the precise location anywhere within a document of any particular set of characters, whereas "replace" is the key to automatic (selective or universal) substitution of one text string for another. The latter facility is especially useful when the need arises to revise a frequently recurring piece of terminology (replacing "Section" by "Sec.", for example), but it can also serve as a shortcut for the entry of cumbersome words or phrases that appear repeatedly within a given document. Thus, the name of a complicated chemical compound might initially be introduced in an abbreviated form ("#1", for example), to be spelled out correctly at a later stage with the aid of a replacement operation.[5]

Equally valuable for the desktop publisher is a more sophisticated "search" or "search-and-replace" option sensitive to specific *formats*, making it possible, for example, to change selectively certain text fragments set

---

[5] Some programs (e.g., WORD; cf. Sec. 9.4) provide elaborate "glossary" functions for accomplishing the same end, as described later in this section.

originally in 12-point Times bold to 10-point Helvetica italic, or to increase the vertical offset of all superscripts from 3 points to 5 points.[6]

● Flexible formatting

Most word processors today offer a wide range of options and tools related to text *formatting*, including right, left, and full justification as well as text centering with respect to any specified set of margins. Also important is *line-spacing* flexibility, including some way of ensuring constant line spacing even in the presence of oversized or vertically offset characters.

● Typefaces and type styles

Access to a variety of high-quality typefaces in several different type styles (roman, bold, italic, etc.) and various sizes, together with an adequate range of special symbols (Greek letters, mathematical symbols), must be regarded as absolutely essential for the modern scientist/writer. The computer display screen should also provide reasonably accurate approximations of the way such characters will later appear in print (WYSIWYG; cf. Sec. 6.3).

● Tables

Especially for scientific applications one should insist on a word processor that features convenient facilities for the creation of tables, including the possibility of straightforward "fine-tuning" during the course of table preparation (to accommodate unexpectedly wide entries, for example). Correct placement of numerical data is accomplished most readily with the aid of *decimal tabs* that cause columns of numbers to be aligned accurately with respect to a decimal point.

● Footnotes

Every well-designed word processor today can be expected to include provisions for the management of footnotes, usually with an option for collecting the notes instead either at the end of each section or of the document as a whole (as "endnotes"). Footnote numbering and renumbering should be automatic (but subject to manual override), shielding the author from complications that otherwise attend the insertion (or deletion) of note

---

[6] This feature was long absent from WORD (Sec. 9.4), but it became available with MACINTOSH Version 5.0.

references in the course of editing, or required changes in the sequence of a set of notes.[7]

It is important to recognize that numerical footnote references are quite different from ordinary numerical characters, since they also exercise a distinct control function, serving as the link between a specific point in the body of a document and a related fragment of auxiliary text. In the event that a particular piece of text needs to be referred to a second time (as in the case of a repeated literature reference), the second indicator must be introduced manually as a pure *number* (equivalent to the note number already assigned automatically), *not* as an automatically assigned (and therefore new!) note number.[8]

● Pagination and running heads

Optional automatic assignment of page numbers—with a wide choice of pagination formats—and some provision for the incorporation of headers and footers (e.g., *running heads*; cf. Sec. 2.9) are convenience features that are now available in most modern word processors.

● Glossary entries

A good word-processing program should permit one to save complex expressions and frequently recurring text fragments in the form of *glossary entries* subject to recall and incorporation whenever and wherever they are required by the issuance of a simple keystroke command. Such glossary entries should also be subject to precise format specification, and the inclusion of graphics may even be possible. Many scientists particularly appreciate the way glossary entries simplify the preparation of manuscripts filled with awkward, frequently encountered technical terms, such as the names of chemical compounds or biological species. A glossary entry can also serve as a convenient shortcut for introducing a letterhead, which in the case of the latest version of Microsoft's WORD, for example, might even automatically incorporate the current date!

---

[7] Automatic numbering is extremely convenient, but it can also prove a source of frustration in the case of text that will later be transferred into a page-layout environment (cf. Sec. 10.2).

[8] An additional word of caution is also necessary at this point: purely numerical (secondary) references of the type suggested would *not* be subject to automatic revision if the corresponding footnote number were to change as a result of text reorganization.

● Spelling and grammar checkers; thesaurus functions

Tools designed to improve the *quality* of one's writing are increasingly coming to be regarded as essential parts of a full-featured word-processing package. Nearly every program now includes facilities for detecting spelling errors, with customized "user dictionaries" assuming responsibility for verifying the spelling of obscure technical terms or recurring proper names. Some spell-check routines are designed to supply instantaneous feedback as soon as a word appears to have been typed incorrectly, but most require the user to initiate a separate verification step—which might be limited to an individual word or a particular text fragment, but could also encompass a complete document. Figure 9–3 illustrates the way spelling errors are reported by the two programs WORD and WRITENOW.

a)
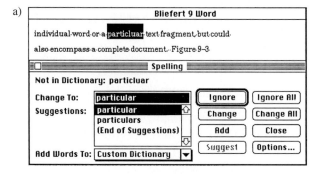

b)
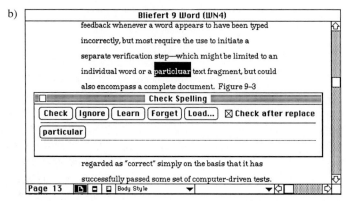

Fig. 9–3. Use of (a) MICROSOFT WORD and (b) WRITENOW to check a document for spelling errors.

A spelling checker plays a useful role at all stages of document preparation, but no document should ever be regarded as "correct" simply because
it has successfully met a series of computer-driven tests. A spelling checker
cannot be expected to recognize, for example, that the word "there" in a particular context should really have been "their" (though errors of this type
can sometimes be detected by a *grammar checker*; see below), nor will it
always locate capitalization or punctuation errors.

Grammar checkers are a more recent innovation, and WORD 5.0 for the
MACINTOSH includes a grammar checker as one of its standard features. A
grammar checker attempts to analyze text according to its structure, relying
on a series of algorithms to single out implausible, undesirable, or blatantly
incorrect word combinations. Unfortunately, language is by its very nature
extremely complex, and most of the available algorithms are incapable of
dealing with sophisticated or subtle constructions. Experienced and capable
writers are likely to discover that use of a grammar checker produces more
aggravation than insight (due to persistent flagging of perceived "errors" that
are in fact perfectly correct), though the writer with less highly developed
language skills may profit (and hopefully learn) from the exercise.

Many writers depend rather heavily on one other type of computer-based
assistance: the *electronic thesaurus (synonym dictionary)*. This is a tool
designed to assist one in selecting precisely the "right" word for a particular
context, or perhaps circumventing excessive repetition. A thesaurus in book
form has long been regarded as a "must" for the serious writer (cf. Ebel,
Bliefert, and Russey, 1987, p. 358), and its electronic counterpart is equally
powerful and at the same time more convenient to use. An electronic thesaurus is now an integral part of WORD (and other programs as well).[9]

● Hyphenation routines

End-of-line hyphenation plays an indispensable role in the formatting of fully-
justified text, and even ragged-right formatting may require occasional use
of hyphens to prevent certain lines from being too short. Hyphens introduced
for justification purposes should always be of the "soft" or "optional" vari-

---

[9] Each of the special functions described here is also available in the form of
subsidiary software, often in desk-accessory format in the case of the MACINTOSH.
Special mention is in order for the AMERICAN HERITAGE ELECTRONIC DICTIONARY
DA (WordStar), which puts all the features of both a standard dictionary (including
word origins, pronunciation guides, and definitions) and a comprehensive thesaurus
conveniently at the user's fingertips.

ety so they will disappear without a trace if they are no longer required. Soft (discretionary) hyphens are normally introduced automatically by program-directed hyphenation routines, but they can be accessed manually as well (usually with the keystroke combination "command-hyphen" in the case of a MACINTOSH, "ctrl-hyphen" under WINDOWS).

Most hyphenation routines operate on the basis of a special dictionary, typically coupled with a set of algorithms applicable to situations not covered by the dictionary itself. The results obtained with such algorithms may be more or less satisfactory depending upon the sophistication of the programming. Like spelling checkers, hyphenation programs are not infallible, and automatically hyphenated text should always be checked with care. (When in doubt, consult a dictionary!)[10]

● Mail merge

A "mail merge" function can be useful in the preparation of form letters and similar documents in which small, predictable changes are expected to distinguish one copy from the next. The first step is creation of a master "boilerplate" document incorporating special markings to indicate words or phrases that are subject to variation. A second document is then created containing the several possibilities for each piece of variable text, after which the word-processing program is instructed to "merge" the two in appropriate ways. A control file provides appropriate detailed instructions, typically qualified by the presence of IF... THEN... ELSE... statements.

One final observation is perhaps in order regarding the selection of a particular word-processing program. Just as with a computer, the most important criterion is a pragmatic one: the intended user should be capable of taking full advantage of all the potential inherent in the product. There is no such thing as a single "best" program, because different people have different needs and different working styles. The most that can be said is that one particular program may be most appropriate in a particular circumstance because it

● promises optimum results with minimum effort,
● offers important features not provided by the competition,

---

[10] An example of a nearly insurmountable hyphenation problem would be the question of whether, in a particular instance, the word "produce" has been used as a verb ("pro-duce") or as a noun ("prod-uce").

- is uniquely capable of conveniently generating text in an unusual format specified by someone else (a publisher or a printer), or
- operates especially smoothly in the context of important auxiliary software.

Our own experience strongly suggests that word processing in general is greatly facilitated—especially for the beginner, but also for the experienced user—by a software system featuring a graphic interface (Sec. 4.2), "pulldown" menus, and mouse control (cf. Sec. 4.4). That said, we hasten to add that the experienced writer already firmly in command of a less convenient word processor will probably prove to be at least as productive as a colleague with a more "user-friendly" system, and that exceptional convenience from the standpoint of the beginner and maximum efficiency for the expert user may well be mutually contradictory requirements.

9.3 Special Considerations with Respect to Scientific Text

Complicated *formulas* (mathematical, physical, and chemical) play a decisive role in many scientific documents, and they often impose demands that extend beyond the scope of an ordinary word processor. The problem might in principle be dealt with in either of two ways:

- One could elect to work exclusively with a special "scientific" word-processing system equipped with integral facilities for the creation and manipulation of all necessary formulas.

Specific programs have in fact been developed to address precisely these needs.[11] Most are structured in such a way as to place heavy emphasis on the preparation of mathematical expressions. The best-known examples are variants of a system known as T_EX (cf. Sec. 10.5), including a Macintosh

---

[11] Special-purpose programs for scientific writers are most common in the MS-DOS world. Examples include T^3, Scientex, and WI-TEX. For an early comprehensive overview of scientific word-processors see C. K. Gerson and R. A. Love in "Technical Word Processors for Scientific Writers", *Analytical Chemistry* **1987**, *59*, 1031A.

program called MacT$_E$X, and the files that result are accepted quite willingly by many scientific publishers. Unfortunately, programs of this type tend to fall well outside the category of "user-friendly" software, and they are not well adapted to the needs of the casual user. Indeed, most bear a marked resemblance to professional typesetting systems, dependent upon extensive sets of coded functions that must be mastered in much the same way one learns to work with a programming language.

- The alternative is to accept the limitations imposed by a general-purpose word processor—preferably one that has already met the test of broad user acceptance—and then rely on auxiliary software for the preparation of essential "special" elements.

The latter approach is probably the best in most cases unless one anticipates routinely spending a great deal of time preparing final copy characterized by a heavy emphasis on mathematical expressions.

If the decision is in fact made to work with a standard word processor, the program selected should be capable of dealing satisfactorily with as many as possible of the following technical elements:

- Subscripts and superscripts

A good word processor for scientific applications must of course provide convenient facilities for introducing both subscripts and superscripts *(indices)*, but also the flexibility that permits individual indices to be adjusted with respect to their size and vertical displacement relative to the baseline.

- Special symbols

Special symbols are extremely common in scientific text, especially letters of the Greek alphabet, so a wide range of symbols should be both readily accessible and subject to expression in satisfactory boldface and italic modifications. Provision should also exist for reasonably faithful screen display of all symbols, together with high-quality hardcopy output.

- Automatic footnote management

The subject of footnotes has already been addressed, but it is particularly important with respect to scholarly manuscripts, including scientific papers, in which considerable emphasis is placed on full documentation of an author's sources. In addition to the considerations presented earlier it is essential that means be available for producing references that correspond precisely with style guidelines established by the editor to whom the manuscript will be submitted, and that the opportunity exists for specifying note placement either at the bottoms of relevant pages or at the very end of a document.[12]

● Automatic preparation of a table of contents

A computer-assisted table-of-contents generator is of greatest interest to the author of a book-length scientific manuscript, or a shorter document with a particularly complicated scheme of subdivisions. The table-of-contents function is designed to produce an accurate list (supplemented with relevant page numbers) of all the section headings present in a given document. Taking advantage of this feature requires that the headings themselves be clearly identifiable, usually accomplished on the basis of unique *style* assignments (cf. Sec. 9.4.1).

● Indexing

Analogous to automatic preparation of a table of contents, automated indexing presupposes that appropriate index terms have been subjected to some type of prior (invisible) marking, either in conjunction with initial entry of the text or as part of a subsequent editing step. Automatic indexing works particularly well with a manuscript characterized by large amounts of technical terminology.

● Extensive facilities for the incorporation of data originating elsewhere[13]

---

[12] Certain word-processing programs include special provisions for conforming to literature-citation demands developed by specific academic societies and publishers. For example, MANUSCRIPT MANAGER: CBE STYLE from Pergamon Press was explicitly designed to facilitate the preparation of scientific manuscripts based on guidelines from the Council of Biological Editors, which are in turn adhered to by the editors of more than 6000 technical journals (e.g., *BioScience* and the *Journal of Clinical Investigation*). The program in fact supports several different CBE citation schemes, including a name–date system (similar to the one utilized in this book), sequential numerical citation, and citations numbered on the basis of an alphabetized list of principal authors. An alternative approach to reference management takes advantage of auxiliary programs like those described in Section 12.6, which are designed to interact with an existing word-processing file in such a way as to introduce *retroactively* a correct set of citation numbers along with the requisite bibliographic data, all on the basis of a separately maintained bibliographic database.

[13] This demand would once have made use of a MACINTOSH computer system almost mandatory, but the rapid development of WINDOWS applications has resulted in an analogous level of compatibility in the PC world (cf. Sections 8.4 and 9.6).

Flexible and convenient data-import facilities are essential for the efficient incorporation of graphic elements into an electronic manuscript: chemical structures, diagrams of various types, mathematical expressions, and even small details like reaction arrows or custom-created symbols. Data import may also represent the most efficient approach to incorporating certain types of textual information, such as complete literature references or names and addresses of colleagues, which might in turn be drawn from external files created with the aid of database or literature-management programs.

9.4 MICROSOFT WORD

9.4.1 Functions

Most of the criteria outlined above could be met by a number of word-processing programs, but perhaps the most satisfactory program, at least from the scientist's point of view, is the highly regarded WORD from Microsoft. WORD in its various stages of development has long been a "best-seller" for the MACINTOSH—indeed, almost since the "Mac" first appeared on the market—and it is also popular among PC users, especially in its WINDOWS incarnation. In this section we describe briefly some of the program's most important characteristics, with illustrations taken from the MACINTOSH version 5.1. The basic set of WORD menus is depicted in Fig. 9–4.

The greatest advantage of WORD over several of its MACINTOSH competitors (e.g. MACWRITE or WRITENOW) is its enormous flexibility and the features it provides for dealing with long and complex manuscripts. MACINTOSH versions as early as WORD 3.01[14] included so many formatting

---

[14] Since December 1985 Apple has recommended uniform use of a "four-place" code to indicate the state of development of a particular piece of software. Only the first two digits (e.g., "3.1") are normally applicable to a product already on the market, although some programming houses add a third digit, as in "3.01", to indicate a very minor revision (Apple prefers to see such a third digit preceded by a second decimal point, as in "System 6.0.7"). Major changes are signaled by an increase in the value of the first digit, those that are less fundamental by changes in the second. A pre-release version of a program (one that is still undergoing testing) acquires a four-place identification of the form "4.1b3", where both the letter and the final number can be expected to increase as the product advances toward maturity.

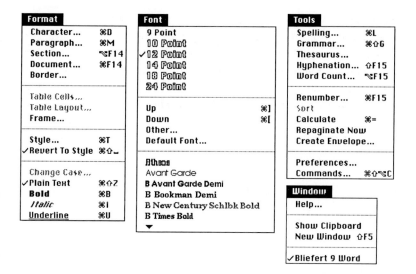

Fig. 9–4. The complete set of default menus for MICROSOFT WORD 5.1.

options that the program was often cited in discussions actually directed toward the more exacting demands of page-makeup software (cf. Chap. 10). Indeed, this one program is capable of meeting nearly all of a technical writer's needs, from preparation of a simple memorandum to the development of a complete camera-ready book manuscript.

Unfortunately, any program as sophisticated as WORD is also bound to be rather complex.[15] Microsoft has attempted to assist the beginner in this respect by minimizing the number of commands and options presented initially in the program's menus, and then offering powerful tools for customizing the menus to suit a particular user's unique needs. As suggested in Sec. 9.2, the best way to learn to use WORD (or any other word processor) is simply to begin typing and experimenting with the most obvious features as the occasion arises (though one should also spend some time at the outset examining the attractive introductory pamphlet entitled *Getting Started*).

As shown in Fig. 9–5, the full WORD screen display incorporates not only a standard menu bar, but also three rows of "buttons". The one at the top is known as the "toolbar", the bottom one is actually part of the ruler, and the one in the middle is called the "ribbon"; display of all three sets of supplementary buttons is optional. A "ruler" is provided for conveniently specifying (and ascertaining) the placement of margins and tab stops.

The most important point to bear in mind when working with WORD is that many format characteristics are defined at the *paragraph* level. Thus, ad-

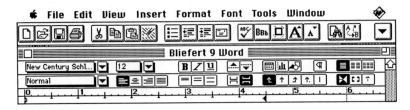

Fig. 9–5. "Buttons" provided in MICROSOFT WORD for more convenient activation of many of the programs's various functions.

---

[15] The same applies to the corresponding manual(s). It is refreshing to be able to report, however, that the WORD 5 manual is a distinct improvement over its predecessors, but the average user would still be well-advised to invest in a carefully prepared third-party tutorial/guide to the program as a supplement to the encyclopedic documentation provided by Microsoft. One of the best books in this category is *The Word 5.1 Companion* by the Cobb Group Staff (1993).

justing a particular setting on the ruler normally results in changes only in the particular paragraph which at that moment contains the text cursor. Instituting a more far-reaching change requires *selection* of a larger section of text (leading to white-on-black display of the text in question), accomplished most simply by "dragging" the mouse through the relevant text while the mouse button is depressed. Whenever a *new* paragraph is created—by pressing the "Return" key—it automatically acquires all the formatting attributes applicable to its predecessor (though each of these characteristics is of course subject to subsequent modification).

One of the most powerful (but often ignored) formatting tools available in WORD is the potential for conferring upon a particular set of paragraph characteristics a unique label known as a *style name*. Thus, the label "body text" might be defined in such a way that it specified a paragraph with one-inch left and right margins, a first line indented by one-half inch, a tab in the center of each line, text set in 12-point Times roman type, and lines separated by four points of extra leading. Whenever a paragraph with this set of characteristics is required anywhere within the current document it can be created instantly by suitably positioning the cursor and then selecting the style "body text" from the list of all available (user-defined and default) styles shown in a "pop-up" list at the left of the screen, directly above the ruler (cf. Fig. 9–6).[16] In a similar way, style characteristics might be established for *headings* at various levels, thereby ensuring format consistency throughout a manuscript without the need for remembering all the corresponding "rules". Once paragraphs have been associated with particular styles, a *change* made in any of the styles at the style-sheet level (in response to an editor's suggestions, for example) would be conferred automatically on all paragraphs bearing the appropriate designation, irrespective of when these paragraphs might have been typed.

Tables can be introduced into a WORD document either in the conventional way with the aid of tab stops (taking advantage of decimal tabs as appropriate) or by calling upon a special "Table Function" that produces a starter set of "cells" (as in Fig. 9–7) analogous to those in a spreadsheet (cf. Sec. 12.4). Each cell in such an array behaves as if it were a separate paragraph, and is thus subject to its own "ruler" settings. Like paragraphs (see

---

[16] Style information incorporated into a WORD document is retained when that document is later "imported" into PAGEMAKER, which also provides facilities for assigning styles (cf. Sec. 10.2).

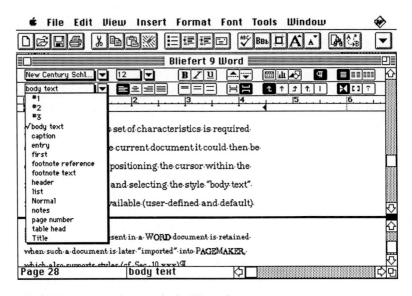

Fig. 9–6. Style menu for a particular WORD document.

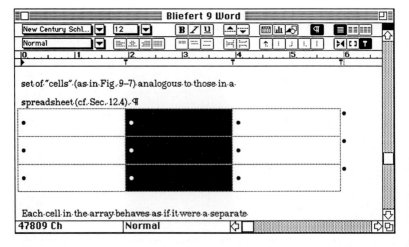

Fig. 9–7. Template for a 3 × 3 table in a WORD document. Note that the ruler displays the boundaries for the various columns, and all cells in the middle column are currently "selected".

below), cells can be selectively surrounded (or partially surrounded) by *borders* of various types, a feature that facilitates the incorporation of dividing lines into a table (cf. Sec. 2.12).

Word documents are normally saved in a file format unique to that program, but several other formats are available as well, ranging from plain ASCII text to the formats employed by MacWrite and the PC programs Word for MS-DOS, Word for Windows, and WordPerfect. This flexibility permits a Word document to be exported into a wide range of alternative environments, including page-layout and spreadsheet programs. Another translation facility simplifies the *import* of text developed elsewhere—even text embedded in a graphics file. Indeed, Word is capable of accessing text elements in virtually *any* file, including the PostScript instructions present in an EPS file. This is the type of flexibility that makes it possible for a complete desktop-publishing operation to be established on the basis of a single relatively inexpensive home computer and a limited set of software.

Finally, Word provides the user with a convenient "print preview" function for seeing in advance how a document will look once it is printed, with special tools that change graphically such global parameters as placement of headers and footers and the extent of the margins (cf. Fig. 9–8).

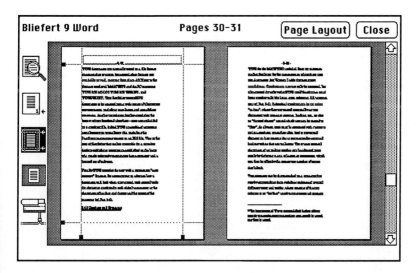

Fig. 9–8. Two pages of a Word document as displayed in "Print Preview" mode.

9.4.2 Graphics and Equations

WORD for the MACINTOSH included from its inception a modest set of facilities for the inclusion of graphic elements in text documents, but Version 5 added a wealth of new possibilities. Now graphics can not only be imported into a WORD document, they can also be created within WORD itself thanks to the presence of a (limited) set of graphic tools (for lines, arcs, polygons, fill patterns, etc.; cf. Fig. 9–9). Individual graphic elements can be set either directly in the text ("in-line"), where they are treated as essentially equivalent to text characters with respect to spacing, leading, and vertical displacement, or they can appear as "framed objects" around which the text can be caused to "flow". Special care must be exercised with respect to in-line graphics, since they may lead to unwanted changes in line spacing due to automatically assigned leading values that are too large.[17] The desired vertical placement of an in-line graphic is achieved most easily by defining the image as a subscript or superscript, after which it can be offset by precisely the appropriate number of points (see below).

Limited text passages can be highlighted in a semi-graphic way by surrounding them with *lines* or *boxes* of several different types and widths.

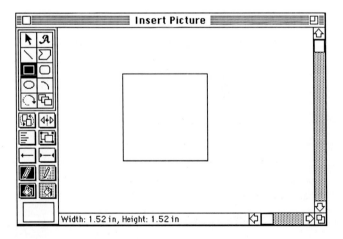

Fig. 9–9. Window for preparing simple graphic images in WORD 5.1.

---

[17] The latest version of WORD supports default line spacing of three types for paragraphs containing graphics: *auto*, *exactly x points*, or *at least x points*.

Spacing of the corresponding text relative to its "borders" is subject to precise independent adjustment on all four sides. *Shading* can be applied to paragraphs as well, and at any desired level. These particular features are all accessed through a "Border…" command in the "Format" menu (Fig. 9–10).

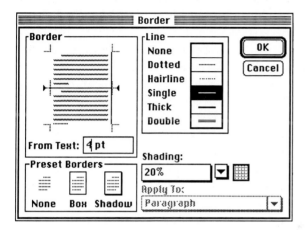

Fig. 9–10. The "Border" dialogue box in WORD.

Particularly important with respect to scientific text is the fact that individual indices (subscripts, superscripts) in WORD are subject to vertical adjustment to the exact level required, and the sizes of the indices are adjustable as well, opening the way to relatively straightforward preparation of professional-looking formulas; e.g.:

$$X_A, X_A, X_A, X_A, ..., X^B, X^B, X^B, X^B,$$

WORD also provides its own unique (but cumbersome!) system for constructing complex mathematical expressions based on a coded form of input, with results that can also be displayed on the screen in WYSIWYG fashion. Thus, "text" entered initially as

$$\text{\I\ISU}(n \cdot = \cdot 1, \infty, \text{\F}(\text{\R}(3n), n^2 \cdot - \cdot 1))$$

can be caused to appear on the screen as

$$\sum_{n=1}^{\infty} \frac{\sqrt{3n}}{n^2 - 1}$$

leading ultimately to very satisfactory printed output from a laser printer. Nevertheless, most users wishing to incorporate numerous complex mathematical expressions into their text would be well advised to ignore WORD's "formula function" and take advantage instead of one of the convenient accessory programs developed explicitly for that purpose (e.g., EXPRESSIONIST, MACEQUATION, MATHTYPE, MATHWRITER; cf. Sec. 9.7), particularly since WORD-built formulas are not subject to extensive "tweaking", and they cannot be imported successfully into other programs like PAGEMAKER. Microsoft is obviously aware of the shortcomings of its text-based formula system, since the company has chosen to include a somewhat limited version of the MATHTYPE equation-editing accessory as part of the MACINTOSH WORD 5.0 program package, complete with a separate user's manual. Unlike earlier versions, WORD 5 takes full advantage of alignment information present in mathematical formulas prepared with EXPRESSIONIST, so an EXPRESSIONIST "picture" imported as an in-line graphic is automatically positioned to conform to the correct text baseline.

9.5 Other Word-Processing Programs

The first word-processing program released for the Macintosh was Apple's own MACWRITE. Originally provided cost-free as a supplement to the standard system software, this early MACWRITE—in some respects a truly revolutionary program—would necessarily be judged as quite primitive according to today's standards. A linear descendant of that venerable program is now distributed by Claris under the name MACWRITE II. Figure 9–11 illustrates the MACWRITE II user-interface. As with WORD, various text formats are available at the click of a button through controls displayed in conjunction with the ruler. MACWRITE originally differed from WORD in the fact that formats were established not at the paragraph level but rather by special "rulers" inserted directly into the text, but this unusual approach has been abandoned in MACWRITE II.

Other MACINTOSH word processors include MACAUTHOR (Icon Technology), WRITENOW (WordStar International), FULLWRITE PROFESSIONAL (Borland International), and WORDPERFECT FOR MACINTOSH (WordPerfect Inc., a major supplier of MS-DOS word processors), some of which are more

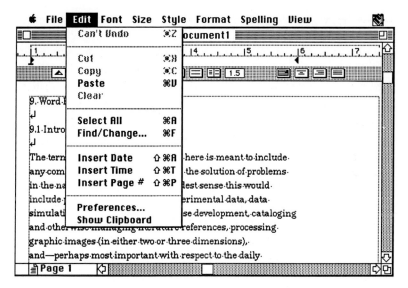

Fig. 9–11. User interface for MACWRITE II.

sophisticated than others. We have elected to forego detailed descriptions of these programs, since they offer few significant advantages over WORD, which in our judgment should more than fulfill the needs of the average scientific writer.[18] One notable exception is WRITENOW, distinguished primarily by its speed and very modest storage requirements, which makes it ideal for POWERBOOK users (for a comprehensive review see the June 1994 issue of *Macworld*).

One other program with a strong word-processing component warrants mention as well: RAGTIME, developed originally in Europe and currently distributed in the United States by MacVonk of Canada. RAGTIME combines powerful word-processing features, an impressive set of page-layout functions, and certain characteristics normally associated only with a spreadsheet program (see Sec. 10.4 for additional information on RAGTIME). Comprehensive program packages like MICROSOFT WORKS and CLARIS WORKS represent other attempts at meeting a variety of user needs simultaneously,

---

[18] A relatively recent comparison and evaluation of several MACINTOSH word-processing programs, including Word 5.0, is provided by T. Landea in the article "The Right Word Processor", *MacUser*, September 1992, pp. 101–105.

but their word-processing capabilities are too restricted to be of interest to the serious desktop publisher.

Yet another interesting piece of word-processor-related software is DocuComp II (Advanced Software), which makes it possible to identify differences between two saved versions of a given text document. Changes, deletions, and transpositions are all detected, highlighted, and summarized for screen display or optional hardcopy output. An early version of DocuComp was at one time distributed as part of the Microsoft Word program package.

9.6   File Formats for Text

The wide variety of existing file formats for text means that not all word-processor files are mutually compatible. This potential limitation is mitigated somewhat by the fact that most word processors do offer facilities for *saving* files in alternative ways, or for *opening* files created by other programs. For this reason many of the general observations in Sec. 8.4 regarding graphic file formats are equally applicable to text files (cf. also the comments on file types in Sections 1.1.2–1.1.4).

Among the most important formats for text files are:

● ASCII, TEXT, "text only"

These three terms are essentially equivalent, and applicable to files consisting exclusively of strings of ASCII characters uninterrupted by coded formatting information. This is the way text is normally treated when it is transferred from one file to another via the Macintosh (or Windows) clipboard, which explains why attributes like "Times 12-point bold" often disappear in the course of a "cut-and-paste" operation.

● DCA-RFT (Document Contents Architecture–Revisable Form Text)

This format was developed by IBM as a standard for its own word-processing systems, but it is of some interest to users seeking compatibility within a network environment.

● MACINTOSH WORD, WORD FOR DOS, WORD FOR WINDOWS

Microsoft markets special versions of WORD for use within both the MACIN-TOSH and MS-DOS worlds, all of which are rather closely related. For example, MACINTOSH WORD 5.1 is capable of reading and writing files in both of the PC formats, permitting facile transfer—with full formatting retained—of text information across the sometimes troublesome Mac/PC boundary.

● MACWRITE, WORDPERFECT

These formats apply specifically to text files created with the programs MACWRITE and WORDPERFECT (cf. Sec. 9.5) or other software compatible with them. WORD 5 is again capable of reading and writing files of both types.

● RTF (Rich Text Format)

RTF format is a Microsoft innovation designed to simplify the transfer of formatted text from one environment to another. An RTF file includes a relatively straightforward and standardized set of codes conveying various types of format information. For example, if the following phrase were to be expressed in 12-point Helvetica:

H_2SO_4 is a *dibasic* acid.

the data would be rendered in RTF as:

```
{\plain \f21 H}{\f21\fs16\dn4 2}{\plain \f21
SO}{\f21\fs16\dn4 4}{\plain \f21 is a } {\plain
\i\f21 dibasic}{\plain \f21 acid.\par}
```

Here the code "\i" indicates italic type, whereas "\f21" specifies the particular font Helvetica.

● SYLK

SYLK is a format used primarily for spreadsheet files created by the Microsoft programs MULTIPLAN and EXCEL (cf. Sec. 12.4).

9.7 Mathematical and Physical Formulas and Equations

Word-processing programs generally offer rather limited sets of tools for developing complex mathematical expressions. For this reason it is advantageous to supplement a word processor with one of several auxiliary programs designed explicitly for working with mathematical equations. The four most widely distributed for the MACINTOSH are:

- MACEQUATION (Software for Recognition Technologies)

- EXPRESSIONIST (Prescience), an easily mastered package that includes both a full program and a powerful desk accessory

- MATHWRITER (Brooks/Cole), described by its authors as "an intelligent full-screen mathematical editor"

- MATHTYPE (Design Science, Inc.), a limited version of which is included with WORD 5 under the name EQUATION EDITOR

The fact that both MACEQUATION and EXPRESSIONIST exist in "desk accessory" versions is convenient when one wishes to move rapidly between text- and equation-editing tasks without the additional memory burden associated with two open programs (under System 7 or MULTIFINDER). MATHWRITER is not available as a DA, and its interface is somewhat more cumbersome, though it does offer the advantage of a special submenu for cataloguing and recalling previously prepared expressions. MATHTYPE is unique in providing its own special typeface for characters not available in the standard Symbol font (cf. Fig. 9–12).

Expressions prepared with the aid of one of these special editors are subject to direct incorporation into most word-processor text files, preferably in the form of "in-line" graphics that retain their proper positions even if surrounding text is moved or otherwise modified. All are inherently capable of generating expressions of roughly equivalent quality, but many regard EXPRESSIONIST as the easiest to use. Figure 9–13 shows an EXPRESSIONIST editing window containing part of the second formula on p. 89.

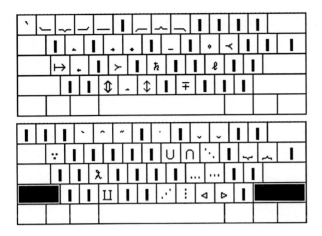

Fig. 9–12. Symbols provided in the special font accompanying the WORD EQUATION EDITOR, shown in relation to a standard MACINTOSH keyboard without (above) and with (below) the shift key depressed.

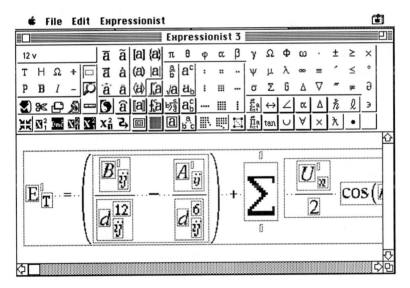

Fig. 9–13. EXPRESSIONIST window displaying a portion of the "improved" equation presented on p. 89.

9.8 General Guidelines for Writing with a Word Processor

Writing with a word processor need not be very different from any other kind of writing—simply more convenient. Most authors will find that they work best by composing directly at the keyboard, with minimal attention in the early stages to niceties of formatting. Once a significant amount of text has been entered it is ready for editing, and this is where the computer-assisted process begins to shine. Some people routinely edit their work on the screen, rephrasing, deleting, and rearranging text elements as necessary to produce more lucid and organized prose than that which developed spontaneously. Others (the present authors included) find it much more productive—and restful—first to generate a printed copy of the crude text, which can then be edited comfortably and systematically with a pencil. Changes introduced in this way are then transferred to the computer file to provide the basis for another clean printout—which is much easier to refine further than a manuscript already filled with editorial marks. Word 5 offers one unique feature that is especially handy at the editing stage: selected text (one or more words, or even several paragraphs) can be "dragged" with the mouse cursor from one location to another as a convenient alternative to a conventional "cut-and-paste" operation.

Most word processors provide the option of viewing text exactly as it will be printed or supplemented with markers for all non-printing characters. The latter form of display (apparent in all the illustrations from WORD in this chapter) should always be selected for editing purposes, since it may otherwise be difficult to recognize unambiguously the presence of extra blank spaces, or the precise locations of paragraph breaks; errors resulting from such ambiguity can sometimes be very confusing.

The following additional suggestions apply to text preparation with word-processing programs of all types:

● Documents should be formulated initially in as concise and systematic a way as possible. Whenever possible, long or complex documents should be subdivided into smaller units (e.g., < 30 KByte). One advantage of this approach is that it leads to more rapid document processing (e.g., file storage), but it also facilitates the subsequent task of transferring the results into a page-layout program (cf. Sec. 10.2).

- The typeface selected for document preparation should be clearly legible on the display screen. If necessary the typeface used initially can always be changed to one more suitable for final printing—but only *after* the document's content, style, spelling, and punctuation have been carefully optimized.

- Early drafts of a manuscript should always be prepared with generous margins and extra leading to facilitate editing.

- Serious thought should be given to preparing an outline prior to the actual writing step. Some word processors (including MICROSOFT WORD) offer special integrated outlining tools to encourage this more systematic approach to creative writing (cf. also the discussion of MORE in Sec. 1.6).

10 Page-Layout Programs

10.1 Introduction

Chapter 9 emphasized the fact that the chief concern in selecting a word-processing program should be convenience with respect to the initial preparation of material consisting largely of text. The most important factors relate to data input, rearrangement and other modification of text once it has been entered, footnote management, assignment of "styles" to confer functional distinctions on specific paragraphs, and other similar matters (cf. Sec. 9.2). Detailed *formatting* of the text and its merger it in an aesthetically pleasing way with graphics and other independent text elements to produce a single complex document with a near-professional appearance is a task better left to a *page-layout* (or *page-makeup*) program. Thus, all the individual subunits that are expected to constitute a final document—especially text, but also illustrative matter of several types—should be prepared in advance with the aid of whatever specialized software seems most appropriate to the task, and then *imported* into a new and permanent setting, after which the pieces can be carefully distributed over the various pages, perhaps subjected to a limited amount of final editing, and adjusted in subtle ways to achieve the desired aesthetic effect.

A good page-layout program must be capable of meeting the following basic requirements:

● Broad access should be available to both *text* and *graphic* files of several different types, consistent with use of a wide range of word processors and graphics applications (cf. Chapters 9 and 11). The import procedure itself should be rapid, straightforward, and totally reliable. In particular it should ensure the retention of all formatting characteristics already conferred upon a piece of text (especially character formatting), as well as faithful incorporation of all graphic elements.

● Provisions should exist for imposing rigorous constraints on *placement* and *dimensions* with respect to both text blocks and graphic elements regardless of their origin.

● The program should facilitate the design and utilization of sophisticated *master* (or *default*) *pages* containing all the elements required for consistent assembly of the basic page units that will characterize a particular document. Appropriate margins, decorative or functional lines and borders, text-column boundaries (for both single- and multiple-column layouts), pagination, running heads, non-printing guide lines to simplify correct placement of text blocks or illustrations, and even repetitive graphic elements should automatically become a part of each new page the instant it is created. At the same time one should also be able to override the default parameters in special situations. Provision for *dual* master pages is also important, since left and right facing pages in a publication must usually be treated in different ways. It is this feature that makes it possible, for example, to ensure that page numbers always appear at the *outside* edge of a page, and that inside and outside margins differ in width. The great advantage of master pages is that they eliminate the burden of repeatedly measuring and marking successive pages as a publication grows (and always remembering to apply the relevant standards!). Master pages thus constitute the key to maintaining integrity of design throughout a document.

● A good layout program should permit the inclusion of *lines* of various widths and styles, often as free-standing elements but in some cases linked in such a way that they produce functional and attractive *borders*.

● Standard *document templates* are useful accompaniments to a basic page-layout package. A template is a skeleton document designed primarily for immediate use in some routine application, but templates can also serve as a valuable source of inspiration for those with little experience in document design.

● Needless to say, the *screen representation* of a "document-in-progress" should closely resemble the corresponding printed output (WYSIWYG; cf. Sec. 6.3).

Quite apart from page-layout features, a modern application in this category should also be outfitted with convenient tools for *editing* imported text, and for creating entirely new text elements. Every text element—whether imported or newly created—must be subject to rigorous typographic specifications. Among the most important are:

- an unlimited choice of typefaces, in all the common styles and a wide range of sizes, including the freedom to intermingle type of various kinds

- sensitive line-spacing parameters that are both adjustable and open to precise control

- flexible kerning facilities (cf. Sec. 2.3)

- powerful word- and letter-spacing algorithms to assist in optimizing the appearance of a block of text, whether justified or unjustified (cf. Sec. 2.6)

- "text wrap" capability, so that a text element can be caused to "flow" around a fixed graphic element (though this feature is rarely utilized in conjunction with scientific text)

- reliable (automatic) end-of-line hyphenation, an essential consideration if one hopes to achieve well-spaced justified text

Layout programs generally fall into one of two basic categories depending upon whether they are conceived around "text blocks" or "frames". With a program organized according to the *block* (also called *column* or *page*) principle, such as PAGEMAKER, pre-set column boundaries and custom guide lines allow one to "pour" a previously prepared text file directly into a blank document. Imported text that exceeds the space available on a single page can be accommodated by establishing *links* to suitable blocks on subsequent pages. Alternatively, the program might be instructed simply to proceed on its own with automatic (default) placement of lengthy text files, making use of the requisite number of immediately succeeding (blank) pages ("auto-flow"). "Live" connections are retained between the various blocks constituting a continued story, permitting text to flow freely back and forth between individual segments in response to editorial changes or layout modifications, mandated, for example, by the subsequent introduction of additional text or graphic elements.

A *frame-oriented* program (e.g., QUARKXPRESS, READY,SET,GO!, RAG-TIME, or FRAMEMAKER) imposes a more rigid segregation of text and graphic elements, in that the user must specify in advance precise locations and sizes for appropriate "frames" (or "boxes") to accommodate the various elements that will constitute a document, all prior to the introduction of the corresponding content. The frame approach emphasizes the important role of document *design*, and for that reason it is particularly well suited to the assembly of a complex publication like a newspaper or magazine, where stylistic considerations and relationships between individual elements are often paramount. However, it also requires that one know from the outset approximately how much space each projected story or illustration is likely to consume.

The sections that follow describe in some detail programs representative of both of the fundamental page-layout categories.[1]

10.2 PAGEMAKER

PAGEMAKER is certainly the most familiar example of a page-based layout system, and it is generally regarded as the most easily mastered of the "mainstream" programs. PAGEMAKER also classes as one of the pioneering MACINTOSH applications, having played a very major role in establishing the MACINTOSH reputation for quality in the field of DTP.[2] The program has essentially retained its original character throughout the course of an extensive evolution, but its flexibility has increased dramatically. PAGEMAKER's grip on

---

[1] A useful overview of a wide variety of MACINTOSH page-layout programs—from entry-level packages to long-document software—is provided in the article "Desktop Publishing Diversifies" by Jim Heid in the February 1992 issue of *Macworld* (p. 208).

[2] PAGEMAKER for the MACINTOSH was first introduced in 1985 by the Aldus Corporation, which recently announced plans to merge with Adobe Systems. Kleper (1990) describes PAGEMAKER as "the grandfather of all desktop publishing programs" and "the program that most effectively has defined and shaped the character of desktop publishing...the linchpin that held together the MACINTOSH and the LASERWRITER as a single-station electronic publishing system." PAGEMAKER screen shots reproduced in this chapter were taken with MACINTOSH Version 5.0, released in 1993. One of the best sources of detailed information on the use of PAGEMAKER for the MACINTOSH is the book by Kvern and Roth (1992); the same authors (1993) have also prepared a similar overview of the WINDOWS version of the program.

the overall DTP market was further strengthened with the release several years ago of a MAC-compatible MS-DOS version.

PAGEMAKER is particularly well suited to the preparation of medium-sized documents containing limited numbers of such special elements as formulas or illustrations. The user interface of PAGEMAKER (cf. Fig. 10–1) includes the usual menu bar and tool palettes, but the bottom of the screen is distinguished by icons representing a set of several consecutive pages (or page pairs) within the current document ("publication"), where the icon for the currently displayed page(s) is highlighted in black. Similar symbols, marked "L" and "R" for left and right, provide access to the corresponding master pages. "Clicking" with the mouse on one of the page icons calls immediately to the screen an accurate representation of all the elements currently constituting that particular page (or page pair), each in its proper place. Any such element is subject to relocation or size adjustment— independently of other elements that may be present—by the issuance of appropriate mouse or keyboard commands. Display scales range from four-fold enlargement to a "fit-in-universe" view (now called "Show pasteboard")

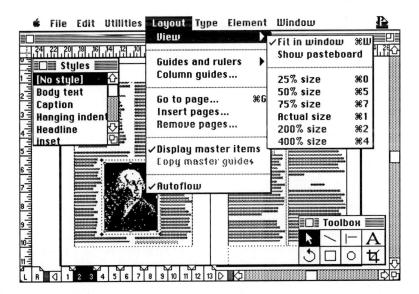

Fig. 10–1. The user interface for PAGEMAKER 5.0, showing a reduced view of pages two and three of a rather long document.

that encompasses the extensive "work surface" surrounding any actual document (see below).

A given PAGEMAKER data file can be saved either as a *publication* (i.e., a unique document) or as a *template*. The latter functions as a type of "dummy" for conferring upon future documents a specific set of pre-determined characteristics, including default text and graphic elements. Unlike a publication, a template is not subject to direct permanent change, because activating ("opening") any template in fact leads to a *new* file that must in turn be saved under its own unique name. Templates are especially useful for situations in which several documents are expected to share a common layout: the various chapters of a book, for example, or successive issues of a serial publication.

As already noted, any page in a PAGEMAKER document can be displayed at several levels of magnification. The user is encouraged to think of the current page (or page pair) as occupying the center of a larger "work surface" that can serve many of the same functions as a traditional layout table. Thus, various elements awaiting placement within a document might be distributed temporarily around the free area at the edge of the screen, ready to be moved into permanent position with a stroke of the mouse. Such "stored" items are completely independent of the page display, so they remain visible even if a different page is called to the center of the "table".

The first step in preparing a document with PAGEMAKER is to establish the general nature of a typical page. This includes setting the *dimensions*—chosen, perhaps, from among the many default options (US letter-size, legal-size, A4, etc.), or specified directly in inches, millimeters, or picas. One is also expected to declare at this point whether the pages in question are to be oriented in "landscape" (wide) or "portrait" (tall) format (cf. Fig. 10–2), after which an initial estimate is provided of the required number of pages. A default *column arrangement* is then established (up to ten columns per page, with variable widths),[3] and guide lines might be introduced to assist in the routine placement of text and graphic elements, as might such recurring features as pagination marks and decorative or structural lines.

All the effort to this point has been directed toward the master pages that will serve as a basis for the real pages constituting the document itself. The next step is to display the first "working page" and then call up for placement

---

[3] Predefined columns are subject to alteration graphically during subsequent stages of document preparation, and pages can also be added or deleted as necessary.

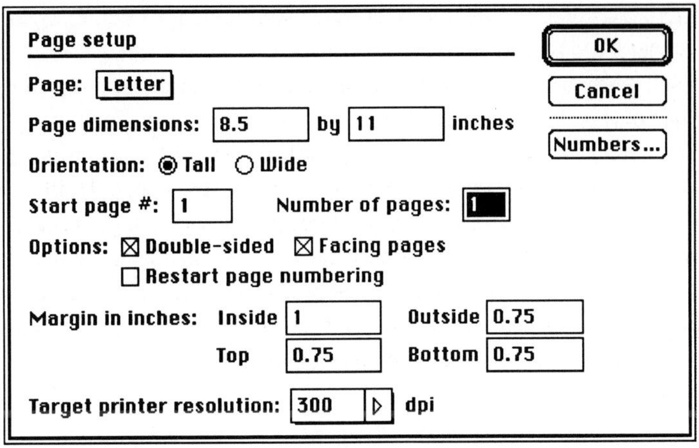

```
┌─────────────────────────────────────────────────────────┐
│  Page setup                              ┌──────────┐     │
│                                          │    OK    │     │
│  Page: │Letter│                          └──────────┘     │
│  Page dimensions: │8.5    │ by │11    │ inches  ┌────────┐│
│  Orientation: ◉ Tall  ○ Wide                    │Numbers…││
│                                                 └────────┘│
│  Start page #: │1    │    Number of pages: │1    │        │
│  Options: ⊠ Double-sided  ⊠ Facing pages                  │
│           □ Restart page numbering                        │
│  Margin in inches:  Inside │1    │   Outside │0.75  │      │
│                       Top  │0.75 │   Bottom  │0.75  │      │
│  Target printer resolution: │300   │ ▷ │ dpi             │
└─────────────────────────────────────────────────────────┘
```

Fig. 10–2. Dialog box used for establishing the initial configuration of a PAGE-MAKER document.

(from an appropriate word-processor file) the first piece of text. This is again accomplished through a dialog box (accessed with the "Place" command in the "File" menu), resulting in a special mouse cursor used for "aiming" the imported text at a specific page location (a particular column, for example). One might of course wish to reserve space in advance for illustrations and other graphic elements, but this can also be attended to at a later stage in the makeup process. Once a piece of text has been "placed", additional character formatting is applied as necessary (e.g., alternative type styles or sizes), although formatting already present in the original text file is faithfully retained (with the exception of margin settings). "Style sheet" tags are usually imported along with the actual text (from WORD documents, for example; cf. Sec. 9.4), but paragraph styles are also subject to assignment or alteration directly within PAGEMAKER itself. Text is adjusted automatically so that it fits within the specified margins, and words at the ends of lines are subjected to whatever hyphenation is required (on the basis of an extensive dictionary and powerful algorithms for dealing with unfamiliar words[4]) consistent with an adjustable set of letter- and word-spacing targets. As might

[4] Foreign-language hyphenation dictionaries are available for separate purchase, and provisions can be made for associating a specific hyphenation dictionary with a given paragraph.

be expected, the program offers an extensive set of tools for fine-tuning the spacing of both words and letters, and manual kerning is available to deal with special problem situations (e.g., the proper alignment of subscript–superscript pairs in complex mathematical or chemical expressions).

Just like a full-featured word processor, PAGEMAKER provides a complete range of tab options, including the decimal tabs so valuable in the preparation of tables.[5] Various borders are also available for highlighting specific paragraphs, as are screen patterns for use as backgrounds. Straight lines of various widths can be drawn anywhere on a page, as can squares, rectangles, circles, and ovals, which can appear either as outlines or as shaded geometric figures. Finally, graphic objects of any degree of complexity, and based on several formats (e.g., PICT, TIFF, EPS, PAINT; cf. Sec. 8.4), can be imported and placed wherever they may be required—either as independent objects or as in-line graphics embedded directly within a block of text such that they move about as the surrounding text is edited. Imported graphics are subject to both proportional and non-proportional distortion, applied either graphically or numerically (e.g., proportional shrinking to 64% of the original size; cf. Fig. 10–3). Text blocks are subject to rotation by any desired amount, and rotated text can even be edited; facilities are also provided for the rotation of graphics.[6]

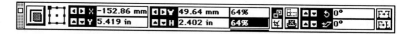

Fig. 10–3. The PAGEMAKER 5.0 "Control Palette", within which parameters can be introduced for changing the size or placement of any element on a particular page.

A text block can be edited directly within the page on which it appears, but it is often more convenient to take advantage of a special text window called the "story editor", which offers many of the text-editing features one

---

[5]   Aldus distributed as part of the PAGEMAKER 4 package a subsidiary program specifically designed for the creation of complex tables, which could then be imported as graphics into any PAGEMAKER document. However, this approach suffered from one serious disadvantage: the graphics themselves were not subject to convenient editing. PAGEMAKER 5 does not include the TABLE EDITOR software.

[6]   This flexibility with respect to rotation is one of the most important features added in Version 5 of PAGEMAKER.

would expect to find in a high-end word processor, including a "search-and-replace" routine sensitive to specific character formats.

PAGEMAKER also includes a feature for obtaining a conceptual overview of a complex publication. Thus, a set of "thumbnails" can be prepared to illustrate a particular range of pages, as illustrated in Fig. 10–4 for the first eleven pages of a lengthy manuscript. Note that even such elements as pagi-

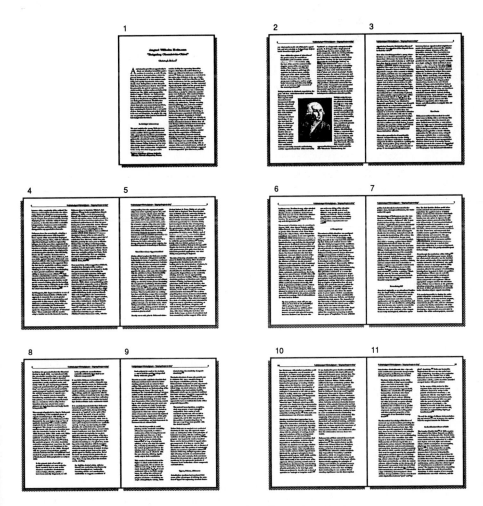

Fig. 10–4. PAGEMAKER "thumbnail" representation of a portion of a document.

nation and headers are clearly distinguishable, though they may not be directly legible. Thumbnails are especially valuable when the time comes to assemble a complex printed document, since a comprehensive set of thumbnails can be regarded as a small-scale reproduction of the complete publication, making it easy to verify correct arrangement of the real pages.

PAGEMAKER acquired relatively recently (Version 4.0) one attribute of considerable importance to the scientific writer: great flexibility with respect to indices. Subscripts and superscripts can now be adjusted individually to conform to almost any desired vertical placement, and they can also be scaled to any required size. Default characteristics for indices are specified in the course of defining paragraph styles. As noted previously (Sec. 9.4.2), equations created on the basis of formula commands in WORD are not interpretable by PAGEMAKER, but mathematical expressions developed within EXPRESSIONIST or MATHWRITER are treated as ordinary graphic objects, and therefore present no problems whatsoever. An equation from EXPRESSIONIST automatically assumes its correct position relative to the text line when introduced as an embedded graphic. Import is also straightforward with respect to chemical equations and formulas created with the aid of CHEMDRAW or CHEMINTOSH (cf. Sec. 11.6).

Several valuable features were added to PAGEMAKER as part of a minor upgrade to Version 4.2. Perhaps most important for the scientist was precise digital control over the placement of text and graphic elements, achieved through a floating "control palette" accessible from the "Windows" menu (cf. Fig. 10–3). Another welcome addition was the ability to continue work on a project without waiting for the screen to be redrawn after some trivial modification of a document. Two important features introduced with Version 5 (albeit at the expense of vastly increased memory and disk-storage requirements) were multiple open-document windows and the ability to alter the sequence of pages. Unfortunately, PAGEMAKER still offers no direct support for footnotes. Footnotes imported as part of a WORD document, for example, are all collected at the end of the corresponding file, from which they must be individually "cut" and then "pasted" (as independent text blocks) on appropriate pages.

10.3 QUARKXPRESS

QUARKXPRESS has long been regarded as the most serious competitor for PAGEMAKER in the world of desktop publishing. Indeed, many users regard QUARKXPRESS as superior to PAGEMAKER, but any such ranking is heavily influenced by the nature of the task at hand as well as the elusive factor of "personal taste". A clear-cut judgment is also complicated by the fact that the two programs have become increasingly similar with time, each acquiring important capabilities introduced originally by the other.[7]

The basic principle underlying QUARKXPRESS and similar programs is that every item on a page is a unique entity defined in part by the (invisible) "frame" with which it is surrounded (cf. Sec. 10.1). Thus, the first step in introducing any element into a document—whether text or graphics—is creating the corresponding frame, which then becomes restricted to objects of a particular *type* (text *or* graphics, cf. Fig. 10–5). As with PAGEMAKER, multiple text frames can be linked to permit the contents of a story to flow freely between blocks: from one page to the next, for example.[8] Graphic elements—which can be surrounded by various types of standard borders, some quite decorative—are normally introduced as freestanding objects. Placing such a graphic in the middle of a text element causes the text simply to move out of the way and "flow" around the intruding object.[9] Graphics can also be embedded or "anchored" within a block of text, although the procedure is somewhat awkward and counterintuitive.

QUARKXPRESS offers much greater versatility with respect to master pages than PAGEMAKER. Thus, a given document can be constructed on the basis of as many as 100 different master pages (in contrast to the two permitted in

---

[7] For rather objective comparative reviews of the two programs see E. Taub, "PageMaker 4.2 and QuarkXPress 3.1", in the July 1992 issue of *MacUser* (p. 56) or the more recent "Professional Page Layout: Time to Switch?" by K. Tinkel and C. England (*MacUser*, February 1994, pp. 108–122), which compares PAGEMAKER 5.0 with QUARKXPRESS 3.2 (and 3.3). The latter article also includes brief discussions of READY,SET,GO! 6.0.2 and FRAMEMAKER 4.0 (cf. Sec. 10.4).
[8] A special provision also exists in QUARKXPRESS for importing long stories in such a way that required extra pages are created automatically without special planning or attention on the part of the user.
[9] A similar feature is available in the most recent versions of PAGEMAKER as well, though with a more limited selection of border elements.

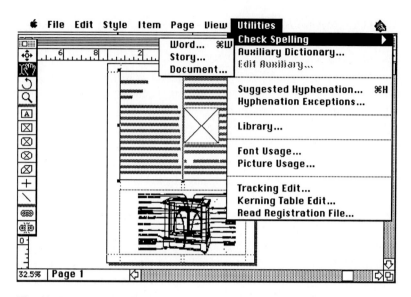

Fig. 10–5. The user interface for QUARKXPRESS 3.1, with "frames" illustrated for both text and graphics (note that one of the graphics frames is still empty).

PAGEMAKER: a "left page" and a "right page"), and a particular master page can be associated at any time with any desired page(s) in the document. Moreover, any element introduced by way of a master page is subject to selective deletion from a page based on that master—unlike master elements in PAGEMAKER, which can only be suppressed as a group (though unwanted features can usually be eliminated by "masking" them with opaque, borderless boxes). Multiple master pages are especially helpful in the preparation of a document with running heads, the content of which is likely to change within the course of a single chapter, or when one wishes to maintain a consistent spatial treatment of illustrations wherever they happen to appear in a document.

Perhaps the greatest strength of QUARKXPRESS lies in the incredible degree of typographical precision it supports. Thus, type sizes in a QUARKXPRESS document can vary over the complete range from 2 to 720 points in increments of 0.001 point! Similarly, leading can be specified to within 0.001 point, and kerning is controllable over increments as small as 0.00005 em. Both text and graphic elements are subject to free rotation in increments of 0.001°, specified either with the mouse or (for greater precision) through a dialog box. One other important feature—strangely lacking in PAGEMAKER— is the power to *group* and *align* objects, facilities now taken for granted in object-based graphics programs (cf. Sec. 11.3) but also useful in a page-layout context. Users concerned with accurate color reproduction report that QUARKXPRESS provides the most sophisticated color-management tools of any available page-layout program.

QUARKXPRESS long offered the very real advantage of permitting access to as many as seven documents simultaneously, allowing one to cut and paste at will between one document and another, but this facility is now a part of PAGEMAKER as well. Both programs support third-party supplementary software to increase basic program functionality (i.e., PAGEMAKER "Additions" and QUARKXPRESS "XTensions"), although the range of available products is broader in the case of QUARKXPRESS.

Major shortcomings of the program relative to PAGEMAKER include a less convenient treatment of style sheets and the absence of facilities for automatic generation of an index or table of contents. Moreover, PAGEMAKER provides a more complete set of drawing tools for the rapid introduction of simple graphic elements. Most beginners also find the PAGEMAKER user interface to be somewhat more intuitive, resulting in a less formidable "learning curve".

Perhaps the best advice for the well-equipped desktop publisher committed to routine preparation of large, demanding documents is to secure access to (and familiarity with!) *both* PAGEMAKER and QUARKXPRESS, selecting one or the other according to the nature of the particular task at hand.[10]

10.4 Other Page-Layout Programs

A second well-established program of the "frame" type is READY,SET,GO!, now distributed by Manhattan Graphics.[11] The user interface for Version 5.14 of READY,SET,GO! is illustrated in Fig. 10–6, which clearly points up the

---

[10] More detailed information regarding the use of QUARKXPRESS is available from such sources as Blatner and Stimely (1991), Blatner and Weibel (1993), or Meehan et al. (1992).

[11] Currently offered as version 6.0.2, early versions of READY,SET,GO! were marketed under the name DESIGNSTUDIO.

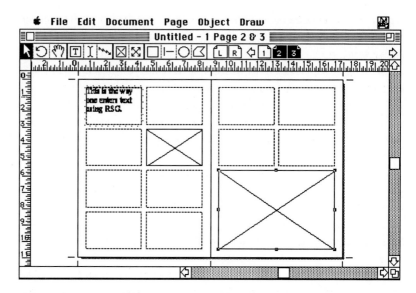

Fig. 10–6. A typical screen display from READY,SET,GO! 5.14, showing the presence of a preconfigured "grid" for consistent alignment of text and graphic elements.

program's unique *flexible grid* system, designed to simplify the preparation of systematic, orderly pages consisting of elements arranged neatly in columns and rows.

As with QUARKXPRESS, one begins the construction of a READY,SET,GO! document by defining frames to house the various text and graphic elements one proposes to incorporate. If it seems appropriate, an orderly arrangement of these frames can be ensured by constraining them such that they automatically "snap" to the pre-existing grid. In contrast to QUARKXPRESS (but like PAGEMAKER), style-sheet information in a WORD document is retained when the corresponding text is imported into READY,SET,GO!. The program also shares PAGEMAKER's double-page-spread display format and basic drawing tools. It includes rather sophisticated tools for manipulating graphic images and color, but still appears better suited generally to the preparation of relatively small documents—ones in which the emphasis is on graphic design—rather than lengthy scholarly treatises. As with QUARKXPRESS, all elements constituting a page are subject to free rotation and enhancement with borders, some with rather imaginative shapes, but important convenience features one normally associates with creating book-length publications (indexing, multiple master pages, etc.) are somewhat restricted.[12]

One unique aspect of READY,SET,GO! worthy of special mention is a very comprehensive *scripting language* ("ReadyScript") to facilitate the execution of long sequences of routine operations or the semi-automated production of groups of documents sharing rather similar specifications.

Perhaps the most sophisticated of all the MACINTOSH frame-oriented page-layout programs for the preparation of lengthy technical documents is FRAMEMAKER (Frame Technology). Less popular among desktop publishers than PAGEMAKER or QUARKXPRESS, it nevertheless boasts an impressive set of features, including extensive WORD-like facilities for the creation and subsequent modification of complex tables and an option for incorporating "conditional text" (passages that vary depending upon what version of a document is to be printed).

Another interesting program, PERSONAL PRESS (distributed by Aldus, which acquired the original developer, Silicon Graphics) is a page-layout package designed explicitly to meet the needs of novices interested in the convenient preparation of small, routine documents (newsletters, brochures, business communications), particularly ones combining extensive graphics with complex page designs.[13] While it lacks the sophisticated typesetting capabilities and flexibility of PAGEMAKER or QUARKXPRESS, PERSONAL PRESS is refreshingly simple to use thanks to a very convenient and intuitive interface. One of the most unique aspects of PERSONALPRESS is the ease with which it allows one to create an entire publication—incorporating several stories and diverse graphic elements—from within a single dialog box customized for the particular type of document required. Here the user is presented with a miniature view of the projected layout together with tools for specifying precisely which files are to be imported into which of the pre-existing frames (cf. Fig. 10–7). The program itself attends to problems of adjusting the content physically so that it conforms to the available space, altering as

---

[12] Version 6 of READY,SET,GO! includes provisions for 26 right- and left-master pages as well as more sophisticated typographic controls and precise mutual anchoring of otherwise independent elements.

[13] Another well-established program in this category is PUBLISHIT! EASY (Timeworks). For a general discussion of low-end page-layout software see "Layouts for Less" by G. Wasson in the July 1991 issue of *MacUser* (p. 110).

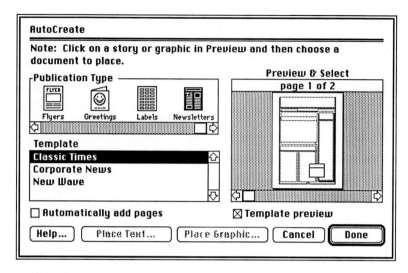

Fig. 10–7. Dialog box for "automatic" preparation of a complex document in PER-
SONAL PRESS (Aldus).

necessary a variety of text-spacing parameters as a way of taking optimal
advantage of the available print area. PERSONALPRESS also supports the use
of a scanner (cf. Chap. 8), so one can capture and incorporate graphic images directly from their source without relying upon auxiliary scan software.

   Probably the most unusual program in the "other" category is RAGTIME
(MacVonk, Canada), currently extolled as a software package optimized to
meet the diverse needs of the business community. The page-layout features
of RAGTIME resemble those of other frame-based programs, but in addition
to frames for text and illustrations this program also offers special frames
for spreadsheets as well as various types of graphs (bar graphs, pie diagrams,
etc.). These can in turn be linked directly to data files, permitting one to
combine in a single publication an astonishing array of fixed and variable
elements. Thus, an invoice form, which would normally consist primarily of
text (perhaps together with a company logo) might be so designed that it
also included a customized spreadsheet frame. The spreadsheet could then
serve as an elegant device for inserting automatically the necessary pricing
information together with a statement of the total amount due, all compiled
on the basis of a separate master stock catalog (in the form of a data file)
and the unique characteristics of the particular order in question (cf. Fig.

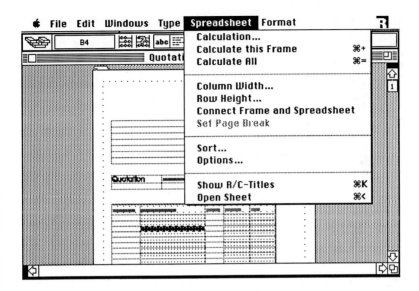

Fig. 10–8. The RAGTIME 3 interface, showing a document containing both text and spreadsheet elements. One cell in one of the spreadsheet elements is shown to be currently "selected".

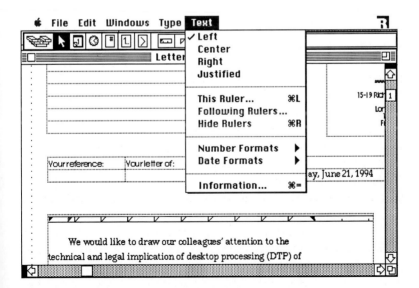

Fig. 10–9. Template for a business letter in RAGTIME 3.

10–8). RAGTIME's spreadsheet capabilities are of course not as extensive as one would expect from a dedicated spreadsheet program like EXCEL (cf. Sec. 12.4), but they are certainly impressive in the context of page-layout software!

RAGTIME also permits one to create templates in which specified elements are "locked", an effective way of preventing unintended or unauthorized changes in a standard form. "Live" automatic date-stamping is available as well, another of the special features that might make this program attractive not only to those from the world of business but also to scientists confronted with a frequent need for constructing forms, or struggling to cope with an unusual amount of routine correspondence (cf. Fig. 10–9).

10.5 Typesetting Software

We conclude this chapter with a brief introduction to *code-oriented* page-layout programs, including TEXTURES[14] (Blue Sky Research, originally from Addison–Wesley), as well as PAGEONE (McCutcheon Graphics), and JUSTTEXT (Knowledge Engineering). Such *typesetting programs* (or *formatting programs*) might also be described as "markup languages" (ML; cf. Sections 1.1.3 and 1.1.4), with the aid of which an unstructured electronic manuscript is converted into a fully-formatted document. This is accomplished by embedding in an existing text file a complex set of formatting instructions expressed in the form of well-defined ASCII strings. The em-

---

[14] This program is based on a well-known typesetting language referred to simply as TEX and devised by Donald Knuth (cf. Knuth, 1987; other sources of information include Bechtolsheim, 1992; Désarménien, 1986; Gurari, 1993; Snow, 1992; and Spivak, 1990). TEX is in effect a complete typesetting system, especially designed to meet the demands imposed by mathematical text—and at a professional level. The TEX language consists of more than 1000 different commands, so it clearly does not lend itself to rapid assimilation. In its raw form it could also not be described as "user-friendly", although some of the programs *based* on TEX do incorporate features that simplify its implementation. In general it is fair to say that TEX is better adapted to semi-professional applications than occasional use by the novice. Several scholarly journals (e.g., those of the American Mathematical Society) are produced on the basis of TEX, and many editors welcome authors' manuscripts submitted in the form of TEX files.

bedded codes then become subject to interpretation by specially programmed *formatters*, which translate them into the instructions required to cause specific symbols to be printed in a particular way at the correct location on a page.

Software in this category reflects the habits and expectations of typesetters accustomed to dealing with professional typesetting equipment. As previously indicated in Section 1.1.3, typesetting codes are introduced directly into the stream of running text. TEX files, and similar files based on other languages (e.g., SGML; cf. Sec. 1.1.4), are completely independent of the device with which they are prepared; in principle, such a file could originate with any word-processing program capable of outputting pure ASCII strings, although special software is often utilized for this purpose instead.

The decision to prepare a manuscript in the form of a TEX file requires that one largely abandon the luxury of working in a WYSIWYG environment.[15] As a result, fundamental principles of manuscript preparation must be mastered in much the same way that one learns a new programming language. There are, nevertheless, certain advantages to be gained, including:

- highly sophisticated and reliable automatic justification and end-of-line hyphenation
- full support for ligatures
- the potential for creating documents of unlimited size, and without restriction in terms of numbers of columns
- automatic creation of a table of contents and index
- consistent and correct typesetting of the most complex mathematical and physical expressions, including optimal placement and correct scaling of all indices

JUSTTEXT is an example of a relatively simple program for control-code typesetting, one that is based on only about 80 distinct commands. JUSTTEXT output is analogous to a coded manuscript prepared according to the rules of the *Chicago Guide* (cf. Section 1.1.3), and the demands imposed upon the author are comparable. Thus, the following code sequence:

---

[15] A few programs in this category do provide special subroutines known as *screen editors* that make it possible to send to the screen reasonable approximations of a final document.

```
{p14}{f4} ... the {f6}square {f4} of
{f6}c{f4}, {f6}c{f4}{ifA}{su2}{f4}{p14}, is
what is of significance here.{ql}{ql}{p14}
{f4}[Co(NH{if3}){if6}]{su2+} ... where
{f6}c{f4}(Cl{if2}) = 0.5 mol/L
```

is required to produce the two lines:

... the *square* of c, $c_A{}^2$, is what is of significance here.
$[Co(NH_3)_6]^{2+}$ ... where $c(Cl_2) = 0.5$ mol/L

Codes introduced in this example convey the following information:

{p14}	Type size; in this case, 14-point
{f4}, {f6}	The font; Times roman, Times italic
{ifX}	Subscript X
{suX}	Superscript X
{ql}	Fixed-space indicator

11 Graphics Programs

11.1 Introduction

Graphics programs are programs that make it possible to use a computer for creating drawings, illustrations, and other useful graphic elements, with progress subject to constant monitoring by reference to a display screen. In other words, the user acquires all the facilities necessary to "paint" or "draw" with commands issued from a mouse or a keyboard.

High-level graphics programs first became accessible to "normal" users with the advent of generous amounts of computer memory, since applications of this type tend to be considerably more memory-intensive than simple word processors. Another important prerequisite was a high-resolution display screen, considered a superfluous luxury in the early days of computing. Invention of the mouse played a role as well, since this opened the way to a fairly efficient and intuitive approach to expressing relative coordinates—certainly more user-friendly than supplying graphic input in the form of keyboard commands. Finally, it was essential that standardized systems be developed for seamless incorporation of the graphics emerging from powerful graphics applications into a common framework that would also accept text-based material derived from a word processor (cf. Sec. 8.4).

"Graphic data-processing"—the digital manipulation of graphics and graphic elements—has evolved at a stunning pace in the past few years. The days when "computer graphics" or "computer-aided design" (CAD) were exotic playthings reserved for the chosen few with access to large, dedicated computers, or to workstation operators associated with major computer centers, now seem very remote indeed. Sophisticated graphics projects can today be undertaken with ease on a personal computer of almost any type.

Generally speaking, graphics programs are of three basic types. The first group consists of bitmap-oriented (or "paint") programs like MACPAINT (cf. Sec. 11.2). Here the final product is a collection of discrete dots or "pixels" (an abbreviation for "picture elements") that the eye subtly merges into a single image. Such an image becomes amenable to detailed editing by magnification of a selected portion (to give what is sometimes referred to as a "fat bits" view), thereby making it easier to add or remove individual dots. Not surprisingly, the dot-matrix technique was quickly adapted for color applications as well, with results sometimes reminiscent of the "pointillist" art of the late 19th century.

Dot-matrix graphics programs have matured to such an extent that they can now be used to create very impressive images, but they suffer from one formidable limitation: enormous memory and storage demands. Thus, even a relatively primitive program like the original MACPAINT, which was limited to a resolution of 72 dots per inch (dpi), equivalent to the resolution of the MACINTOSH display screen, generates roughly 800 dots per square centimeter, or nearly 500 000 dots for an image the size of a sheet of letter paper. While a bitmap graphic of this type might look rather impressive on a computer display screen, the effect produced by a laser printer or other high-resolution output device is rather disappointing (cf. Sec. 11.3). A more sophisticated program like SUPERPAINT (cf. Sec. 11.2) can, in principle, provide graphics with the same 300-dpi resolution as a laser printer, but taking full advantage of this technique results in ca. 2 MBytes of data for a single full-page illustration. This poses problems not only with respect to storage and the time required for printing, but also in terms of manipulating so much data within the computer.

The second approach to graphics is the *object-* or *vector-oriented* program,[1] also known as a *draw* program (e.g., MACDRAW, MACDRAFT; cf. Sec. 11.3). Here the images created are of a type known as *vector graphics*, in which a circle, for example, is defined not as a collection of points, but rather as a single entity that can be expressed mathematically in the form of a set of coordinates and appropriate geometrical parameters. Compared to a bitmap description, information of this type consumes an almost insignificant amount of storage space. A more recent development based on the same vector-graphic principle is the *PostScript-oriented* or *Bézier-curve* program (e.g., CRICKETDRAW, ILLUSTRATOR; cf. Sec. 11.4). Programs in this category gen-

---

[1]   A few graphics programs offer both bitmap *and* vector-graphic capabilities; see below.

erate graphic forms expressed in the PostScript printer language (Sec. 7.6), which makes them subject to immediate hard-copy output.[2]

The third basic category of graphics software encompasses *3D-programs* (e.g., INFINI-D, RAY DREAM DESIGNER; cf. Sec. 11.5), which permit the preparation of images subject to three-dimensional or *spatial* representation and manipulation.

All the programs described in the next four sections could be of potential interest to almost any scientist, and their use presupposes no special graphic training or skill (though artistic talent might well be reflected in the aesthetic quality of the resulting graphic output). In contrast to some CAD programs and other high-level graphics applications they offer the working scientist a convenient way of preparing visual images of almost any type. The chapter concludes with a brief section (Section 11.6) describing graphics programs designed specifically for preparing chemical formulas.

11.2 Bitmap-Oriented Graphics Programs

The pioneering program in the bitmap category, MACPAINT (developed by Apple, but supported in its more recent versions by Claris Corp.), was originally released in 1984 in conjunction with the first MACINTOSH computers, and in the early years it was almost synonymous with MACINTOSH-based graphics. Indeed, MACPAINT can be regarded as the model for an entire generation of graphics software. The program operates exclusively in bitmapped *(pixel-oriented)* mode, generating images in the form of structured arrays of dots that are archived as simple bitmaps. Thus, each defined point constituting part of a larger image is associated with its own bit of data storage, which can be tagged as either "black" or "white". All MACPAINT images are of the same size, consuming equal amounts of storage space (or memory).

---

[2] A high-resolution graphic image might also be expressed in the PostScript language directly, without any help from a graphic display (cf. Sec. 6.4). Ordinarily, however, one would resort to this procedure only in the rare case of an image that would be difficult to prepare in any other way, and was specifically intended for incorporation into a text document.

The program places at the user's disposal a fairly diverse set of graphic "tools", each of which is associated with a unique screen icon (cf. Fig. 11–1). A given tool is selected by "clicking" on its icon with the mouse. Some tools permit one to prepare freehand images of any desired shape, while others are limited to the construction of simple geometric forms like squares, circles, or ovals, which may be left "empty" or filled with any of several patterns (illustrated in the palette toward the left of the screen image in Fig. 11–1). Any section of a MACPAINT image is subject to as much as eight-fold magnification and closer examination by selecting it and choosing "Zoom In" from the "Goodies" menu, leading to an enlarged "fat-bits" representation in which individual pixels can easily be altered (cf. Fig. 11–2). Support is also provided for copying an entire image or any selected portion for subsequent insertion copy into another document—which might itself consist largely of text.

A complete inventory of the tools provided in MACPAINT reveals:

● "brushes" of various shapes and widths

● a "pencil" for drawing fine lines (and for making alterations at the pixel level in "fat-bits" mode)

● an "eraser" for eliminating superfluous graphic elements

● a "spray can" for distributing dots (or even a background pattern) over a large surface area

● a "paint bucket" for "filling" closed curves with patterns

● a line-drawing tool for creating straight lines of various widths

● tools for creating both open and filled geometric forms (rectangles—which may also display rounded corners—together with circles, ellipses, and miscellaneous polygons), as well as irregular closed curves

● a text tool for introducing alphanumeric symbols in any font and size (where the resulting text is subject to further manipulation, much like any other graphic element)

● a "marquee tool" for selecting rectangular segments for additional processing (copying, cutting, physical displacement), as well as a "lasso" for the selection of irregularly shaped regions

● a "hand" capable of moving an image about on the imaginary work surface

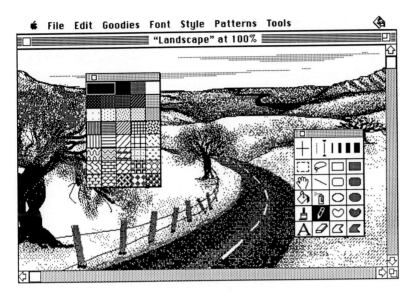

Fig. 11–1. The user interface for MACPAINT 2.0, showing both the pattern palette and the tool palette; the "pencil" tool is currently selected.

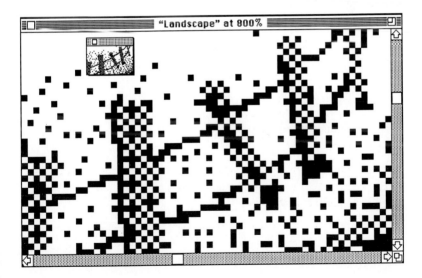

Fig. 11–2. Eight-fold enlargement of a small section of the "painting" in Fig. 11–1, shown at normal scale in the inset at the upper left.

As noted above, MACPAINT documents are limited strictly to a single set of dimensions: 20 cm × 25 cm, only about one-third of which is normally visible on the screen at a given time. A selected portion of such an image can also be displayed (and altered) in an enlarged representation, and at several levels of magnification.

Newer programs of this type, designed to overcome many of the limitations imposed by MACPAINT, include SUPERPAINT (Aldus, formerly Silicon Graphics) and CANVAS (Deneba). Both provide a much wider range of tools as well as stunning color effects, together with many of the "drawing" capabilities described in Sec. 11.3. CANVAS is the more powerful of the two, and it is also available in a WINDOWS version, but CANVAS has been criticized by some as being unnecessarily complicated to master. The user interface for Version 3 of CANVAS is illustrated in Fig. 11–3. CANVAS was actually first introduced as a modest desk accessory, but it quickly outgrew that restrictive format. Sophisticated painting tools continue to be available in inexpensive desk-accessory form from an alternative source, however: Zedcor, the distributor of DESKPAINT (cf. Fig. 4–9).

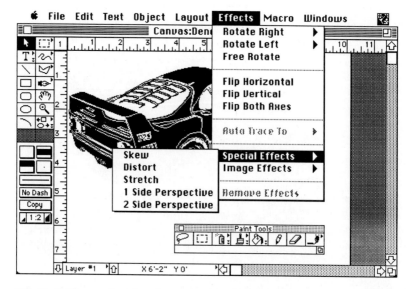

Fig. 11–3. The user interface for version 3.0 of CANVAS. Many of the "tool icons" expand in the form of "pop-up menus" when selected with the mouse.

Mention should also be made of two high-level paint programs designed for use by accomplished (or aspiring) artists: FRACTAL DESIGN PAINTER (Fractal Design) and PIXELPAINT PROFESSIONAL (Super Mac Technology), both of which feature tools for simulating such effects as chalk sketches and work on textured canvas. The most widely distributed paint program for the PC is probably the venerable PCPAINTBRUSH (Z Soft).

11.3 Object-Oriented Graphics Programs

MACDRAW (from Claris) was one of the first home-computer programs for preparing "drawings" rather than "paintings": that is, it was expressly designed for creating graphic images defined *mathematically*. The original MACDRAW was a remarkably versatile software package, equally well suited to both the amateur and the skilled technical artist or architect.[3] It is still perhaps the most familiar example of what has come to be known as *object oriented* or *vector-graphics* software (cf. also Sec. 11.1). In principle, such a program offers the possibility of unlimited resolution, the only practical limits being imposed by the available output device (e.g., a display screen or a printer). The user interfaces for MACDRAW and MACPAINT share many common features (cf, Figures 11-1 and 11–4), but the programs nevertheless represent drastically different (and complementary) approaches to the creation of graphics. In a very few special cases (construction of a hollow square, for example), results from the two might appear identical, but they would still have been achieved electronically in entirely different ways. MACDRAW and its counterparts opened the way to routine preparation of structured graphics of remarkable complexity, as well as images conforming to almost any reasonable set of overall dimensions.

As described in the preceding section, graphics prepared with a program like MACPAINT consist ultimately of collections of dots (pixels), and they are thus strictly analogous to the corresponding screen representations. From the standpoint of the computer, a "painted" circle is not a single entity, but

---

[3] When MACDRAW first appeared on the market, *MacUser* magazine commented (in the issue for September, 1985): "... MACDRAW is about to dramatically expand your horizons. MACDRAW is a serious technical graphics tool ...."

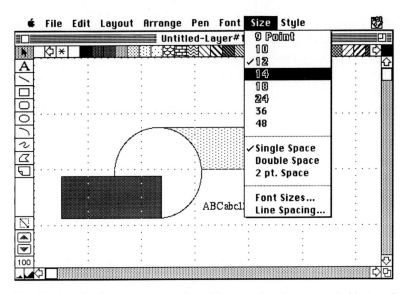

Fig. 11–4. A simple drawing consisting of three overlapping (opaque) objects as it would appear on the screen with MACDRAW 2.0.

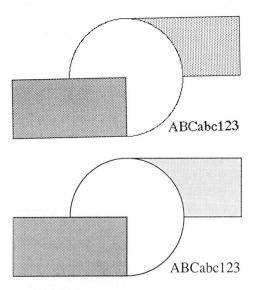

Fig. 11–5. Laser-printer output from the MACDRAW image shown in Fig. 11–4 (below) compared to an equivalent image (above) printed from MACPAINT.

rather a set of tiny dots that the *viewer* perceives to be *arranged* in a circle. This is the fundamental reason why the representation of such a circle cannot be "improved" by substituting for the display screen a high-resolution output device like a laser printer (cf. Fig. 11–5).

On the other hand, a program like MACDRAW stores information about the constituent parts of a "work of art" in terms of the overall characteristics of underlying geometric forms, expressed in QuickDraw format, a far more economical approach to data management. A circle is thus treated as a geometric form with a definite center (associated with a specific set of coordinates) and a unique radius, constructed on the basis of a line with a particular width and perhaps "filled" with some pattern. When a MACDRAW document is transferred to a printer this descriptive information is communicated in such a way that the printer itself acquires the ability to reproduce the desired circle, conferring on it the highest possible degree of perfection (resolution). Even a simple dot-matrix printer like the IMAGEWRITER is capable of expressing graphic output from a MACDRAW image in a "high-resolution" mode with results that are noticeably superior to those achievable with MACPAINT. It is for this reason that users with access to a laser printer are urged to confine their graphic efforts almost exclusively toward software of the DRAW (or PostScript) type.

One important consequence of this alternative approach to graphic definition is the added flexibility it provides. Thus, "paint"-type images cannot be subjected to free rotation or even enlargement without serious degradation in image quality, and most are incapable of profiting from the high resolution of a laser printer or imagesetter. Vector graphics, on the other hand, can be selectively altered in almost unlimited ways with no loss whatsoever in sharpness or definition. Vector graphics are also much easier to modify than bitmap graphics. Furthermore, the "draw" approach permits one to combine several distinct objects in separate "layers", the sequence of which can be changed at will. Thus, a circle might initially be placed "above" ("in front of") a square, causing a portion of the latter to be obscured, and later moved "behind" the square so that part of the circle would instead disappear (cf. Fig. 11–4). Finally, vector graphics print much more rapidly than bitmapped graphics, since it is unnecessary in this case for the status of *every* point on the page to be defined—only those that play a direct role in the visual form in question. Information transfer to the printer also occurs very rapidly, because the extent of the information to be communicated is usually quite limited. Vector graphics lend themselves to output by a plotter as well—perhaps via the intermediacy of a program like MACPLOT (cf. Sec. 6.4.3)—an unthinkable challenge with respect to a bitmap image.

MACDRAW was probably the first personal-computer program to hold out the promise of relatively high-quality CAD-like output at modest cost. Given this breakthrough, there no longer exists a compelling reason for the average scientist to learn the classical techniques of mechanical drawing: better results can usually be achieved more rapidly (even by a beginner!) through use of a computer. Figure 11–6 illustrates the point with a set of symbols for chemical engineering and vacuum technology applications prepared with a MACINTOSH computer according to the exacting specifications of DIN, the *Deutsches Institut für Normung* (1986).

There is one further distinction between MACPAINT and MACDRAW that is important to emphasize. As noted in Sec. 11.2, the "work surface" provided by MACPAINT is very limited in its extent (a consequence of the data-storage requirements of bitmapped graphics). MACDRAW II, on the other hand, supports drawings as large as 2.50 m × 2.50 m—even though few users are so fortunate as to have access to a plotter capable of reproducing such a large image. In most cases a large drawing would actually be output as a set of overlapping "tiles", each the size of a standard sheet of paper.

Every discrete element included as part of a MACDRAW image is subject at any time to selective displacement, distortion, enlargement (or reduction),

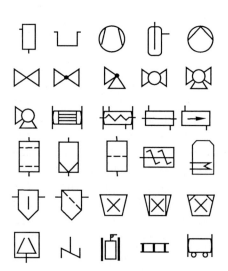

Fig. 11–6. Symbols applicable in chemical engineering and vacuum technology, prepared to DIN standards in MACDRAW.

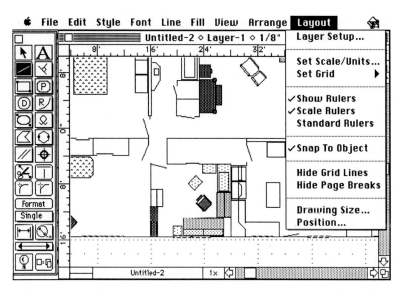

Fig. 11–7. A MacDraft 3.01 scaled representation of a floor plan.

relocation in terms of "layers", duplication, rotation, or inversion. Independent elements can also be "grouped" as a prelude to effecting changes on all members of the group simultaneously, and selected items can be aligned relative to one another in a variety of ways.

Several programs other than MacDraw offer similar capabilities, together with unique features of their own. One worth at least brief mention is MacDraft (from Innovative Data Design), another of the ground-breaking early Macintosh graphic applications. MacDraft is optimized for the preparation of relatively straightforward *scale drawings* conforming to precise specifications. It is thus a valuable tool for developing plans that might be required for later construction of a piece of apparatus, for example, or an accurate representation of an architectural space (cf. Fig. 11–7). Other Macintosh programs featuring a MacDraw-like working environment are IntelliDraw (Aldus) and DeskDraw (Zedcor)—the latter a drawing program in desk-accessory form. SuperPaint and Canvas feature separate graphic "layers" that permit one to combine "painted" and "drawn" elements in a single document. In the IBM world the leading object-oriented graphics program has long been CorelDraw (Corel).

11.4 PostScript-Oriented Graphics Programs

Like the programs described in the preceding section, PostScript-oriented graphics programs rely for their images on defined *objects*. Unlike the former, however, the objects in this case are characterized by the page-description language PostScript (cf. Sec. 7.6) rather than QuickDraw or one of its counterparts. Unfortunately, programs operating on the PostScript principle have difficulty offering the benefits of true WYSIWYG behavior (cf. Sec. 6.3), which is what led someone to coin the acronym "WYSIWHOOPS" (What You See Is What Happens On Operating in PostScript). In other words, such a program must somehow translate the PostScript (printer) commands it generates into a reasonably faithful display-screen representation, in effect serving as a screen-oriented PostScript interpreter.

One program in this category is CA-CRICKETDRAW (Computer Associates), a powerful PostScript-oriented graphics application whose user-interface resembles that of MACDRAW. CRICKETDRAW provides tools for rapidly creating such shapes as rectangles, polygons, ovals, curves, arcs, and freehand lines. A particular line in this case is not actually "drawn", however, but rather "fitted" into a framework defined in terms of certain crucial *points*. The process is heavily dependent upon a concept called the *Bézier curve* (or *Bézier cubic curve*),[4] which bears the name of Pierre Etienne Bézier, the Frenchman who in the 1960s first recognized the potential significance of such constructs with respect to computer-assisted drawing. A bézier curve is essentially a plot of a third-order polynomial, the characteristics of which are established by four points: a starting point, an end point, and two "control points" for specifying curvature. The curve itself is ordinarily created visually by selecting the appropriate "tool" from a tool palette and then clicking with the mouse at the four key spots—in a very specific order. What appears on the screen is a smooth curve joining points one and four but *tangential* to straight lines joining points one and two and points three and four (cf. Fig. 11–8).

A CRICKETDRAW document is of course not limited to geometric figures, straight lines, and bézier curves. Like most other graphics applications the

---

[4]   Certain PostScript-language *operators* also utilize these curves—in the simulation of an arc, for example.

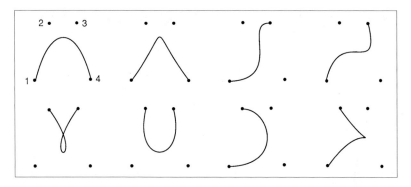

Fig. 11–8. Bézier curves prepared in CRICKETDRAW. All the curves shown are based on the same set of four defining points, but the sequence of point definition has been varied.

program also accepts text blocks with characters set in various styles, fonts, and sizes, analogous to what one might expect from a word processor like MACWRITE. The text capabilities of CRICKETDRAW go far beyond those of a word processor, however, since facilities are also provided for "binding" text to a *curved* path defined on the basis of an invisible line (cf. Fig. 2–18). Images are subject to examination at several levels of magnification, as well as enhancement through a number of special effects, including shadowing, reflection, rotation, and gray-scale patterns. Figure 11–9 is an example of a graphic image prepared in CRICKETDRAW and subsequently output to film by an imagesetter (Sec. 7.4).

CRICKETDRAW is somewhat unique in one respect: it is also equipped to provide the user with a text file containing the set of PostScript commands required to describe an image that has been created graphically. Activating the appropriate program function produces at the right of the screen a new window displaying the PostScript equivalent of whatever image has most recently been "drawn" (cf. Fig. 11–10). These PostScript commands can then be exported to become part of a text document created by a program like WORD (Sec. 9.4). In this sense CRICKETDRAW might be regarded as an example of an easy-to-use and powerful PostScript editor (cf. Sec. 7.6).

Other highly regarded programs for creating sophisticated PostScript graphics include ADOBE ILLUSTRATOR (Adobe Systems) and FREEHAND

Fig. 11–9. A graphic image consisting of text and a gradient fill, prepared with CRICKETDRAW and output with an imagesetter at a resolution of 1693 dpi.

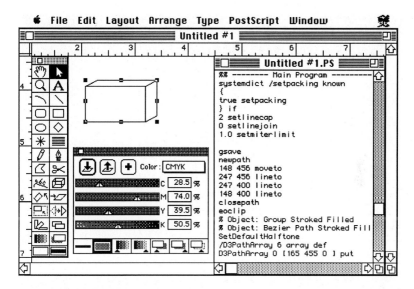

Fig. 11–10. CRICKETDRAW screen simulation of a box (an "extruded rectangle"), together with a scrollable window containing the corresponding PostScript instructions.

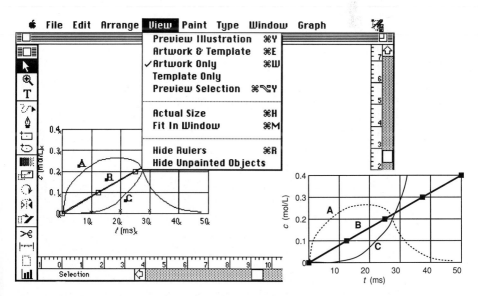

Fig. 11–11. The user interface for ILLUSTRATOR 3.0, showing a graph as it appears in the course of preparation, and the same graph as output by a laser printer.

(Aldus), which exist in both MACINTOSH and WINDOWS versions.[5] Figure 11–11 presents an example of the ADOBE ILLUSTRATOR 3.0 user interface together with both display-screen ("working mode") and laser-printer output of an ILLUSTRATOR image.

11.5 3D-Graphics and CAD Programs

An exciting recent development has been the release of programs that permit even a novice to experiment with the graphic potential inherent in images with true three-dimensional characteristics.[6] Such software puts users of small personal computers in the position of accomplishing feats that until a few years ago were strictly in the realm of massive computer installations featuring powerful CAD facilities.

Programs of this type operate in terms of the three separate dimensions breadth, height, and depth within a spatial framework defined (explicitly or implicitly) by three cartesian axes joined at right angles somewhere in space. Once a three-dimensional object has been "created" it becomes subject to distortion with respect to all three axes as well as free rotation relative to the observer. This makes it possible to view the object or even "illuminate" it from any arbitrary vantage point, and then modify its characteristics as appropriate. Most such programs provide the tools required for creating simple, regular three-dimensional forms directly, including spheres, rectangular solids, and conic sections. These basic elements are then altered and combined in various ways to simulate more complex three-dimensional objects. Alternatively, a two-dimensional form created in another application can be transformed three-dimensionally by "extrusion" or "lathing".

Dealing with three-dimensional images can involve tasks in three categories: *modeling* (establishing the basic form), *rendering* (adjusting various lighting, texture, and surface characteristics), and *animation* (which in-

---

[5] For an extensive comparison of the two programs see "Freehand vs. Illustrator" by Deke McClelland in the June 1994 issue of *Macworld* (pp. 132–140).

[6] For relatively recent reviews of MACINTOSH software in this category see "The Third Dimension" by David Biedny in the September 1992 issue of *MacUser* (pp. 114–130) and "Depth-Defying Design" by Carlos Domingo Martinez in the August 1993 issue of *Macworld* (pp. 92–99).

cludes viewing the image from a new perspective). Available programs differ in the way these several operations are approached and in the degree of flexibility they provide. Among the most popular (and affordable) comprehensive MACINTOSH programs for 3-D work are RAY DREAM DESIGNER (Ray Dream) and INFINI-D (Specular International). DIMENSIONS (Adobe Systems) is a less expensive alternative that offers more limited creative capabilities, but has been optimized for convenient three-dimensional elaboration of artwork prepared separately with a program like ILLUSTRATOR (cf. Sec. 11.4).

A second category of "special" graphics packages is what is known as "CAD" (Computer-Aided Design—or Drafting). Such programs are designed for preparing the formal drawings demanded in the engineering and architectural worlds. CAD programs range from elaborate and expensive packages like AUTOCAD (Autodesk) and VERSACAD (Computer Vision), available for both MACINTOSH and PC systems, to the more modest MACINTOSH program BLUEPRINT (the drafting component of Graphsoft's MINICAD), Autodesk's GENERIC CADD, and WordStar's incredibly inexpensive KEY CAD COMPLETE, with versions for WINDOWS, DOS, and the MACINTOSH.

11.6 Chemical Formulas

Warr (1987) has proposed that a formal distinction be made with respect to four general types of software for dealing with chemical structures in the context of a microcomputer:[7]

● Scientific word processors

● Software with facilities for drawing chemical structures for subsequent incorporation into text, but lacking a suitable basis for carrying out searches based on the structures or their constituent parts

● Software for incorporating into a text environment chemical structures that are also amenable to complete substructure searching, where the structures themselves are expressed in a form compatible with other types of software

---

[7] An excellent overview of programs for drawing chemical structures on several types of computer systems is provided by Meyer, Warr, and Love (1988).

● Software for use in molecular modeling

A related subject (cf. Warr, 1987), the use of vector graphics in conjunction with the storage and recall of molecular diagrams, exceeds the bounds of the present discussion, and will not be considered further. Our intent here is simply to provide an elementary introduction to straightforward structure-graphics possibilities available to the modern chemist, suggest the nature of

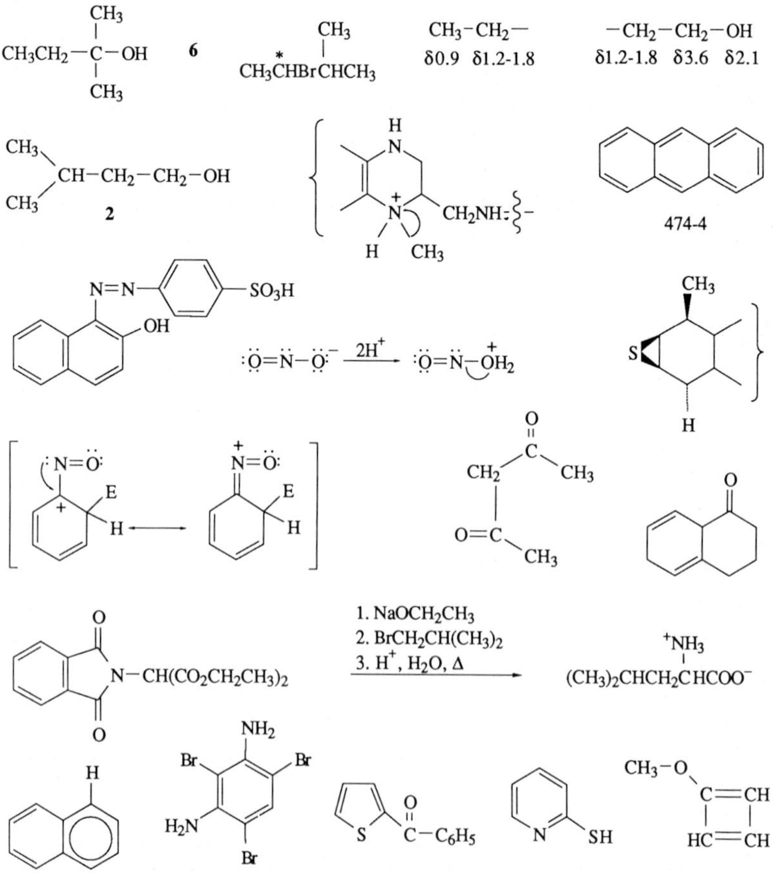

Fig. 11–12. Chemical structures prepared with the MACINTOSH program CHEMDRAW.

the typical user interface, and indicate something about the quality of the results.

Two programs dominate the market for MACINTOSH-compatible chemical structure-drawing software: CHEMDRAW and CHEMINTOSH. We begin with a brief look at CHEMDRAW (Cambridge Scientific Computing, Inc.), a program that leads to structures of remarkably high quality even in very complicated cases (cf. the examples in Fig. 11–12). CHEMDRAW can be described as a good general-purpose program for meeting all the everyday structure-drawing needs of the average chemist.

Launching the program leads to a work surface featuring three menus together with a vertical palette containing icons for the several tools used in creating bonds, rings, and other graphic elements, as well as tools for adding the appropriate text to formulas and equations (cf. Fig. 11–13). Collections of arrows, brackets, and boxes are also accessible through special icons that generate graphic "submenus" for selecting the appropriate symbol (cf. Fig. 11–14).

A great many special functions are provided as well; the user is thus free to select drawing parameters suitable either for large or small reaction

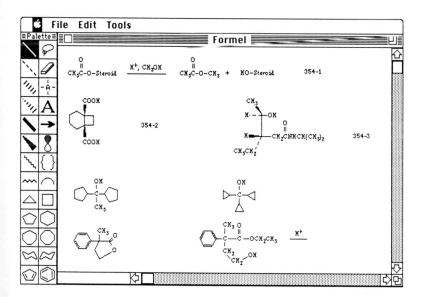

Fig. 11–13. The user interface for CHEMDRAW.

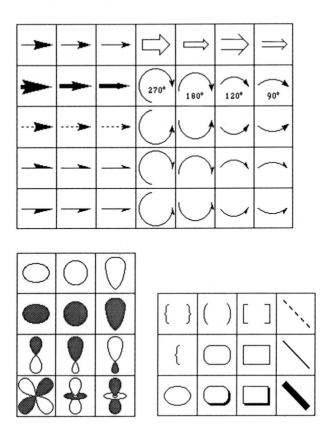

Fig. 11–14. CHEMDRAW menu icons for various types of arrows, orbitals, enclo-
sures, and lines

schemes, and to prepare structures with wide or narrow lines (where line
lengths and widths can be specified in centimeters, points, or inches), as
illustrated by the dialog box that results from selecting the "Preferences"
option in the "File" menu (cf. Fig. 11–15).

Creating the formula designated as (g) in Fig. 11–16, for example, entails
first selecting the six-membered ring tool (fourth from the bottom on the
right; the adjacent five-membered ring tool is not appropriate here since it
leads to structures based on equilateral pentagons) to produce form (a). The
"eraser" tool is then used to remove the top two bonds, leading to (b), after
which the figure is closed to the required five-membered ring (c) with the
single-bond tool. Retracing the two vertical bonds with the single-bond tool

| Preferences | | Cancel | Temporary | Permanent |

Units	○ inches	Chain Angle [◁▨▨120▨▨▷] degrees
	◉ cm	Bond Spacing [◁▨▨15▨▨▷] % of length
	○ points	

		New Captions	Labels
Fixed Length	0.6 cm	Font Times	Font Times
Bold Width	0.141 cm	Size 10	Size 10
Line Width	0.009 cm	☐ Bold ☐ Outline	☐ Bold ☐ Outline
Tolerance	0.141 cm	☐ Italic ☐ Shadow	☐ Italic ☐ Shadow
		☐ Underline ☐ Formula	☐ Underline
Margin Width	0.071 cm	

☐ Fractional Character Widths
☐ Include Footer
☒ Include ChemDraw Laser Prep in Clipboard
☒ Include PostScript commands in Clipboard
☐ Show 35mm Slide Boundary Guides
☒ Fixed Lengths Enabled at Startup

Fig. 11–15. The "Preferences" dialog box from CHEMDRAW.

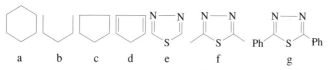

a   b   c   d   e   f   g

Fig. 11–16. An example of the steps involved in creating a complex molecular formula in CHEMDRAW (see text for explanation).

generates the desired double bonds, which automatically acquire the correct length and spacing characteristics (d). The text icon in the fourth row of the tool pallete can be used to introduce the heteroelements nitrogen and sulfur, as shown in (e). The two missing bonds are created as before with the single-bond tool (f), and the structure is completed by typing in the required phenyl substituents with the heteroatom tool (g).

A structure-drawing program of this type offers the great advantage of providing instantaneous feedback through its WYSIWYG character, which greatly facilitates the preparation of molecular representations that are both clear and unambiguous. Analogous programs in the MS-DOS world tended in the past to lack this "reporting" character and corresponding "user-friendliness", meanwhile forcing the user to master an extensive set of command-key combinations more closely aligned to the output device than to the thought patterns of the practicing chemist.

A second program in the same category, CHEMINTOSH (Softshell International) is a more recent entry into the market, one that has made tremendous strides. It shares many of the features of CHEMDRAW, and was the first to offer the added advantage of full compatibility with an MS-DOS (WINDOWS) counterpart, CHEMWINDOWS (also distributed by Softshell). CHEMINTOSH actually began life as a convenient desk accessory, but like CANVAS it soon outgrew this format and was transformed into the full-fledged program now known as CHEMINTOSH 3.3. The user interface for CHEMINTOSH is illustrated in Fig. 11–17. Like CHEMDRAW, the program is capable of producing extremely high-quality output with a minimum amount of effort or preliminary planning (cf. Fig. 3–3). The CHEMINTOSH program package features an extensive library of prefabricated structures and structure fragments, as well as an impressive array of shortcuts for accelerating the pace of structure preparation. An especially useful "accessory" is a sophisticated and reliable routine for checking the chemical plausibility of a structure once it has been prepared, warning the user, for example, of the presence of "pentavalent carbon" (cf. Fig. 11–17).

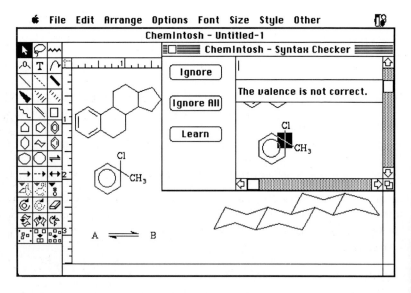

Fig. 11–17. The CHEMINTOSH 3.3 user interface, showing an example of the way a perceived valence error is reported.

12 Miscellaneous Software

12.1 Introduction

Scientists and engineers have discovered a remarkable number of ways to use computers—ranging from direct accumulation of experimental data in conjunction with analytical instruments, or information storage in sophisticated electronic databases, to the processing of numerical quantities in an attempt to transform them into lucid graphic images, charts, or diagrams. It would be impossible (and inappropriate) for us to present here a comprehensive overview of all the specialized software that has been developed to meet the varying needs of scientists generally, especially since the demands of a biochemist, for example, are quite different from those of a process engineer. Nevertheless, certain types of programs do find application in a wide range of scientific fields. We have elected to focus our attention in this final chapter on products of the following types, all of which have the potential to contribute in one way or another to the goals of desktop publishing:

- *Database programs* for the creation and subsequent management of collections of information—addresses, spectra, experimental results, literature references, and many others (Sec. 12.2)

- *Mathematical programs* for analyzing and modifying numerical information, usually for the purpose of casting the material in a more meaningful form, a category that will be regarded as including software for conducting statistical studies and preparing graphs (Sec. 12.3)

- *Spreadsheet programs*, designed to assist in the interpretation and evaluation of tabular data (Sec. 12.4)

- The HYPERCARD environment, a unique object-oriented computer world that can help clarify interrelationships linking various types of data, with an emphasis on graphic display (Sec. 12.5)

● *Miscellaneous software* applicable in a variety of disciplines, as well as powerful *software tools* that facilitate work with the computer itself (Sec. 12.6)

12.2 Database Programs

12.2.1 General Observations

Scientists are confronted on a regular basis with the need to secure rapid access to various types of data. The best way to organize most reference data today from the standpoint of efficient searching is in the form known as a *database*. Typical examples might include comprehensive records of

● one's own publications
● reprints and other bibliographic resources constituting a personal research library
● important addresses and telephone numbers
● information derived from experimental measurements, perhaps accumulated in the context of a research program
● standard spectra and other physical data

A *database program* is essentially nothing more than a set of tools for gaining access to a body of electronic information, including provisions for searching complex data files on the basis of multiple criteria.

The most important factors to consider in selecting a database program for routine use include the ease with which data collections of various types can be established, the breadth of available search options, search speed, and flexibility with respect to data output, including export into alternative software environments (e.g., word-processor files).

Regardless what commercial database package one adopts, it is crucial that implementation be preceded by a considerable amount of time devoted to the design of a suitable *structure* for accommodating the data of interest. Thus, one should attempt to foresee at the outset—as analytically as possible—all the ways the data in question might ultimately be used, because restructuring an existing database in response to unanticipated demands can prove a formidable task. Questions to be addressed at this stage include:

• What is the true nature of the data, and how might the information be expressed most effectively?

• What important structural relationships, if any, serve to link individual pieces of data with one another?

• Which data are clearly essential to the final record, and which might be regarded as superfluous?

• Should certain pieces of information be lifted out of their normal context[1] to facilitate their use as search criteria?

• Are there factors unique to the data in question that must be taken into account to ensure successful recall of the information under various circumstances?

Generally speaking, database programs can be divided into two broad categories as a function of their underlying architecture. A *flat-file (tabular)* database, also known as a *file manager*, organizes information (text, numbers, graphic images) within the context of a single comprehensive data table, from which individual pieces of data or groups of records *(entries)* can be recalled with the aid of an appropriate search protocol. By contrast, a *relational* database has as its underlying principle the linkage of several distinct, highly focussed data files in such a way that redundancy is minimized. In this case the goal is not to assemble an entire complex of loosely interrelated facts into a single table, but rather to analyze the information and distribute it over a series of smaller units, each reflecting a unique vantage point. For example, one subdatabase might be restricted to literature sources relevant to a particular research project (described in terms of principal author, year of publication, keywords, etc.). A second might house information required for maintaining contact with one's professional colleagues (addresses, telephone numbers, key publications, etc.)—some of whom are of course authors of papers included in the literature database. Appropriate links between the two files could thus be exploited whenever the occasion arose to initiate correspondence with an author. Another simple example might be a database designed to serve as a catalog and inventory for expendable materials routinely utilized within one's research group, a database that in the case of chemicals could be linked to a separate file containing physical prop-

---

[1]   A typical example would be the postal codes from a set of addresses.

erties, literature references, and commercial sources. A useful third data file here would be an annotated list of supply houses. If a review of the catalog indicated stocks were running low on a particular chemical, it would need to be reordered—presumably from one of the sources listed (with addresses) in the suppliers catalog. Assuming appropriate formal *connections* exist between the various databases, the relevant information could easily be extracted and combined (perhaps automatically!) into a timely "report", in this case so arranged that it could serve as an order form. Such a structured approach to data management has the potential for not only maximizing information-storage efficiency, but also reducing dramatically the time required for locating and processing data.

12.2.2 Flat-File Database Software

Probably the most popular flat-file database program for the MACINTOSH family of computers is FILEMAKER PRO (Claris), a product that (unlike its Microsoft counterpart FILE) has undergone considerable evolution since its first release in 1985 as FILEMAKER.

The creation of a new FILEMAKER database begins with development of a list of the requisite "data fields", as illustrated in Fig. 12–1. Once all the appropriate field names have been assigned, a *layout* window is used to incorporate the corresponding data fields, one at a time—together with suitable labels—into a *template* reminiscent of a traditional file card. The *location* and *space allocation* associated with each field is established graphically; *formatting* with respect to both field labels and field data (font, type size, type style, etc.) is applied to the template in roughly the same way as with a word processor (cf. Fig. 12–2). The resulting "file-card form" can then be embellished as desired with graphic elements, and might even be supplemented with mouse-sensitive "buttons" to provide users of the file with a simple way of initiating particular courses of action (cf. the discussion of HYPERCARD in Sec. 12.5).

Data are subsequently entered into the file with the aid of this same template by requesting the creation of an appropriate number of "new records", each of which is in turn supplied with the requisite information directly from the keyboard. Data can also be introduced automatically from a preexisting data file, perhaps prepared with a word processor; the only special requirement in this case is that individual data elements be readily distinguishable—

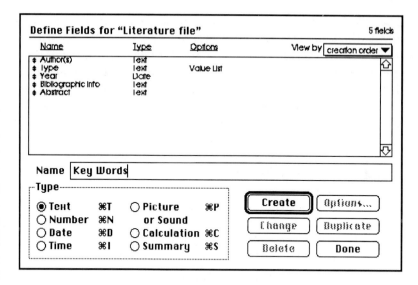

Fig. 12–1. Defining a set of fields for a new FILEMAKER database.

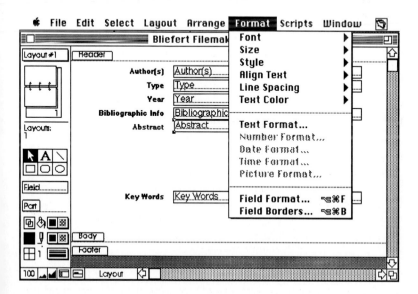

Fig. 12–2. Formatting a template in FILEMAKER PRO.

separated, for example, by "tab" characters. FILEMAKER permits one to create multiple data templates for each data file, where a given template may include all or only some of the data fields. This in turn opens the way to convenient preparation of *summary lists*—or even *mailing labels*—in addition to the usual file-card-like records. Once prepared, a set of FILEMAKER records is easily rearranged into any of several logical sequences with the aid of a "Sort" command capable of dealing simultaneously with multiple sort criteria (cf. Fig. 12–3).

Another example of a MACINTOSH flat-file database program is PANORAMA (ProVUE Development). PANORAMA is somewhat more complicated to master than FILEMAKER, but it rewards the patient user with incredible output flexibility, blinding speed of operation, and an impressive array of shortcuts and safeguards with respect to data entry. The normal data-entry process with PANORAMA is rather analogous to working with a spreadsheet program (cf. Sec. 12.4); indeed, the program has been described by its developers as a "database that thinks it's a spreadsheet", as illustrated in Fig. 12–4. Individual "data cells" in a PANORAMA data sheet can be interrelated with the aid of a wide range of mathematical and logical procedures, and facilities are also provided for fairly sophisticated statistical and graphical treatment of incorporated data. One especially attractive feature of PANORAMA

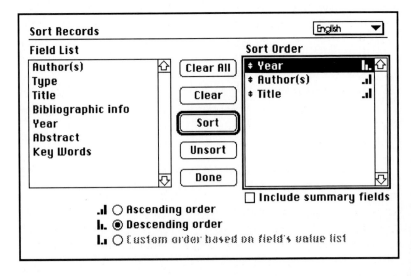

Fig. 12–3. Menu for sorting records in FILEMAKER PRO.

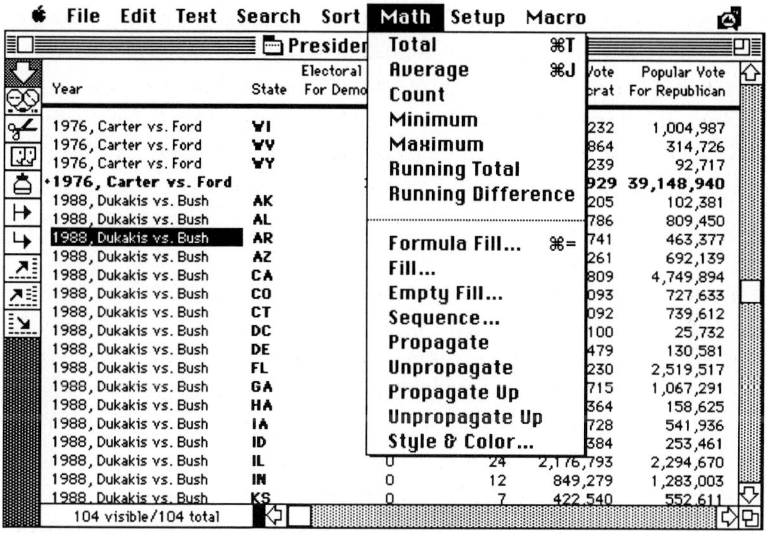

Fig. 12–4. A typical data-entry form in PANORAMA.

is the availability of several routes to simplified data entry, including "Clair-voyance", by means of which the program is empowered to "guess" the nature of input as it is being introduced, by extrapolating on the basis of the first few typed characters. PANORAMA's exceptionally rapid responses are a consequence of the fact that the complete data base is kept constantly in memory (with provision as well for frequent automatic saves), eliminating the delays most programs incur through disk searches. PANORAMA also supports an extraordinary array of options for data import and export, together with limited facilities for linking data files as one step in the direction of a relational database.

12.2.3 Relational Databases

One of the most powerful relational databases available for the MACINTOSH is 4TH Dimension (ACI US), now offered as Version 3.1 in both standard and "professional" packages. Despite its enormous potential, 4TH DIMENSION has been described by many as remarkably user-friendly, in part because of extensive and effective use of icons for accessing program features.

The first step in establishing any relational data bank is to create the appropriate framework: specifying the nature of the proposed set of individual files, naming the respective fields in these files, and indicating ways in which the files are to be regarded as interrelated. Individual data fields must also be characterized according to their content: as text, numbers, dates, symbols, or pictures. Creation of a 4TH DIMENSION data bank is a highly visual process conducted on a work surface resembling that of an object-oriented graphics program (Sec. 11.3). "File folder" icons representing the files themselves can be positioned anywhere one chooses, and connections between the various folders are established with simple strokes of the mouse. A "zoom" function is provided to help one maintain an overview of the emerging structure. A special "programming field" makes it surprisingly easy to access the many tools and functions at the user's disposal, an approach that effectively addresses the problems one would otherwise associate with occasional use of a complex piece of software.

4TH DIMENSION is ideally suited to the person faced with unusually demanding organizational challenges. In contrast to most other complex database systems (e.g., Borland's DBASE for the PC, or the Microsoft equivalent, FoxPro, available in both MACINTOSH and PC versions), 4TH DIMENSION substitutes straightforward mouse operations for the error-prone drudgery of stringing together awkward text-based commands drawn from an obscure database programming language. The search and sort routines in 4TH DIMENSION also represent state-of-the-art technology in both effectiveness and speed. Nevertheless, the average user with a limited range of database needs is probably better served by starting with a relatively intuitive flat-file data-management system of the type described in the preceding section rather than attempting to design a full-fledged relational database.

12.2.4 External Databases

No matter how many custom databases one establishes for everyday use, occasions will still arise that require recourse to external sources of information. This problem, too, is increasingly being dealt with electronically— via personal work stations equipped to access *online databases*. The process usually entails the intermediacy of ordinary telephone lines and an electronic device known as a *modem* (an abbreviation for *mo*dulator— *dem*odulator), although electronic information is sometimes acquired by subscription either in the form of complete data files or as periodic updates, distributed on floppy or CD-ROM disks (cf. Chap. 5). Examples of electronic information available from external sources include:

- Economic and current news databases
- Legal databases
- Industrial databases (e.g., patent information)
- Literature databases
- Chemical databases
- Medical databases
- Environmental databases

In particular, scientific and technical *numerical data* can often be located much more efficiently and rapidly by taking advantage of an "online" (i.e., computer-accessible) database rather than consulting traditional print-based media. Large online database collections are sponsored and maintained by a number of commercial entities, including Dialog Research Services and STN International, as well as by the remarkably informal voluntary organization known as the "Internet" (cf. also Sec. 4.5).

The first step in preparing to access online data is perhaps the easiest: acquiring a suitable modem, preferably one with the potential for transferring data at a relatively high rate of speed: i.e., a "baud rate" of at least 14 400 bits per second (bps). Good equipment can now be purchased for less than $100, often with added facilities for transmitting and receiving electronic facsimile (FAX) documents. The next step is selecting one or more appropriate pieces of telecommunication software, since any program provided "free" with the modem itself will probably prove unsatisfactory. Fortunately, one of the best MACINTOSH telecommunication programs— ZTERM, conceived and written by David Alverson—is readily available from a variety of online sources for a token fee of only $30 (payable on the honor system), an example of what is known as "shareware" (cf. Sec. 4.4).[2]

Telecommunications programs are used for establishing the basic communication parameters that govern conversation between one's own computer and a remote computer that is to serve as the source of target data. An unfortunate complication attending computer telecommunications is the fact

---

[2] Another popular (albeit complicated) "shareware" telecommunications program, RED RYDER, has recently evolved into a commercial product known as WHITE KNIGHT, distributed by a company ironically called FreeSoft.

that different computer systems demand different parameter settings. Most communications packages (including ZTERM) deal with this problem in an almost invisible way by associating the appropriate parameters directly with the telephone number of the computer one wishes to contact. The parameters themselves must initially be entered into dialog boxes analogous to the ones shown in Fig. 12–5.

Once communication problems have been attended to (generally a one-time ordeal), the next step is locating and contacting the database of interest. This is usually a straightforward matter in the case of a commercial service like DIALOG, but it may require a considerable amount of ingenuity and patience with a fragmented "system" like the Internet. We strongly advise the novice to seek help from others who have had more experience, but also to take the time necessary to read one or more of the many Internet guides

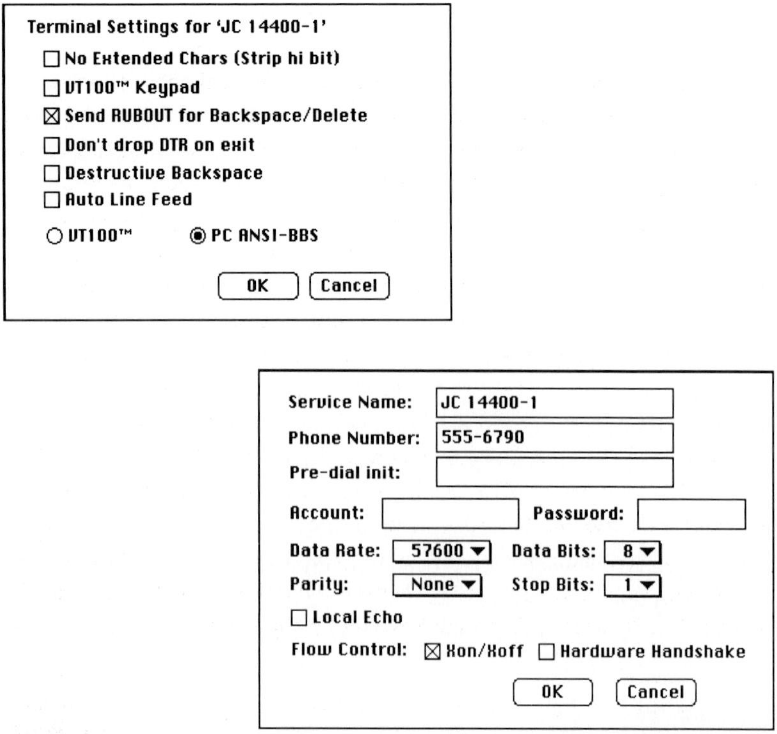

```
┌────────────────────────────────────────────┐
│  Terminal Settings for 'JC 14400-1'          │
│    ☐ No Extended Chars (Strip hi bit)         │
│    ☐ VT100™ Keypad                            │
│    ☒ Send RUBOUT for Backspace/Delete         │
│    ☐ Don't drop DTR on exit                   │
│    ☐ Destructive Backspace                    │
│    ☐ Auto Line Feed                           │
│                                               │
│    ○ VT100™        ● PC ANSI-BBS              │
│                                               │
│              [  OK  ] [ Cancel ]              │
└────────────────────────────────────────────┘
```

```
┌──────────────────────────────────────────────────────┐
│  Service Name:   │ JC 14400-1            │              │
│  Phone Number:   │ 555-6790             │              │
│  Pre-dial init:  │                      │              │
│  Account:  │           │   Password: │           │     │
│  Data Rate:  [ 57600 ▼ ]   Data Bits: [ 8 ▼ ]          │
│  Parity:     [ None ▼ ]    Stop Bits: [ 1 ▼ ]          │
│  ☐ Local Echo                                          │
│  Flow Control:  ☒ Xon/Xoff  ☐ Hardware Handshake       │
│                        [  OK  ] [ Cancel ]             │
└──────────────────────────────────────────────────────┘
```

Fig. 12–5. Dialog boxes for establishing communications parameters in ZTERM.

available in most book stores (e.g., Dern, 1994, Gilster, 1993, Braun, 1994, and especially Hahn and Stout, 1994).

Finally, *searching* an external database once it has been located is an art in its own right. A search query posed in an inappropriate way may only result in failure to elicit a comprehensive response, but it could also produce an unbelievably large number of unintended and irrelevant "hits"—a nuisance under the best of circumstances, but often a costly one as well because of charges levied by the database *host*. Indeed, many organizations have found it advisable to add professional *searchers* to their in-house library staff, individuals specially trained to work closely with scientists in urgent need of specific items of information.[3]

12.3 Mathematical and Statistical Programs

Scientists invariably tend to think first of mathematical applications whenever the subject of computers is raised—quite a natural reaction given the mathematical origins of the very word "computer". Nevertheless, it has only relatively recently become possible for complex mathematical and statistical procedures to be carried out satisfactorily with a small desktop computing device. Applications of this type were long restricted to large mainframe computing systems, the only computers with adequate memory and sufficiently rapid processing ability. The situation has changed dramatically in the past few years, however, and for many of the same reasons that high-level desktop publishing has become a reality (increased RAM, faster processors, etc.). Some of the mathematical and statistical packages now available for small personal computers are astonishingly powerful and sophisticated, and many offer the important additional advantage of "user friendliness", opening the way to their use even by "qualitative" scientists who might otherwise be expected to shy away from esoteric calculations.

The dominant software package at the present time for pure *mathematical* work is MATHEMATICA (Wolfram Research), although the somewhat less for-

---

[3] For more information on the complex subject of searching one should become familiar with such sources as Armstrong and Large (1992), Bottle and Rowland (1992), Harley et al. (1990), and Parker and Turley (1986).

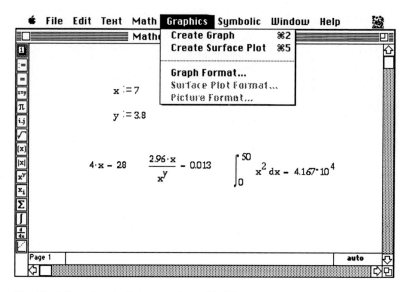

Fig. 12–6. Sample calculations performed in MATHCAD.

midable MATHCAD (MathSoft) represents an attractive alternative. Both require a reasonably fast computer and a substantial amount of memory (at least 4 MBytes) for satisfactory operation. The MATHCAD user interface (MACINTOSH version 3.1) is shown in Fig. 12–6, which also illustrates some rather straightforward, arbitrarily chosen equations that the program was able to solve virtually instantaneously based on the user-defined data appearing at the top of the display. Altering the data set would lead to automatic recomputation of all the related equations. The developers of MATHCAD have made a special effort to be sensitive to the needs of the occasional user, supplementing the program with a comprehensive and exceptionally graphic "help" system (cf. Fig. 12–7).

Statistical packages have experienced a similar trend away from the rarefied atmosphere of the mainframe computer to the desktop of the individual scientist. Examples of widely distributed statistical packages for the MACINTOSH include INSTAT (GraphPad Software) and STATVIEW (Abacus Concepts), as well as SPSS (SPSS, Inc.) and MINITAB (Minitab, Inc.), both very familiar to statisticians as a result of earlier implementations on other hardware platforms. The choice of a particular statistical package is largely a matter of taste, although it may be influenced by the availability of certain

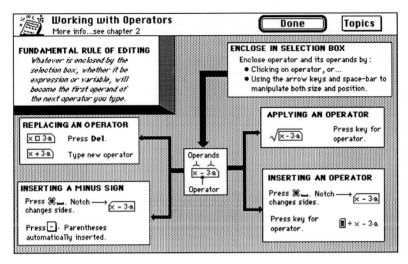

Fig. 12–7. One of the many "help" screens available to users of MATHCAD.

features considered essential in specific situations. It is important at this point for us to reiterate, however, that many modern database programs (e.g., PAN-ORAMA, Sec. 12.2.3) also support the most frequently required statistical procedures, as do some spreadsheet programs (e.g. EXCEL, Sec. 12.4), so a separate statistics package may not always be necessary.

What might be described as a "hybrid species" in this context is a *graphing* (or *charting*) *package* like CA-CRICKETGRAPH (Computer Associates) or DELTAGRAPH PROFESSIONAL (DeltaPoint). Here the emphasis is on versatility and professional quality with respect to data *presentation*, but the presentation features are often supplemented by a fairly wide range of data-manipulation options. For example, Fig. 12–8 illustrates the result of entering a set of numbers into a CA-CRICKETGRAPH data sheet and then subjecting them to the programs' standard set of variable-analysis tests (as if they represented multiple measurements of a single fact). When CRICKET-GRAPH was then asked to take the same "data points", plot them against "row number", establish the best possible fit to a third-order polynomial, and then apply "error bars" to the data points (equivalent in this case to ± 3% of the corresponding values), what emerged was the graph presented as Fig. 12–9. This particular graph is actually a rather simplistic example of the program's true capabilities, as illustrated by the more elaborate demographic chart in Fig. 12–10.

	⌥ File Edit Data Graph Type Arrange View Options					🎲

Data #1

	1	2	3	4	5	6	7
	Column 1		Col C1 Stats	Values	Column 5	Column 6	Column 7
1	27		Minimum	22			
2	29		Maximum	30			
3	28		Sum	189			
4	30		Range	8			
5	22		Mean	27			
6	28		Median	28			
7	25		St. Dev.	2.7080128015453201			
8			Variance	7.3333333333333333			
9			Count	7			
10							
11							
12							
13							
14							
15							

Fig. 12–8. User interface and data-introduction window for CA-CRICKETGRAPH III.

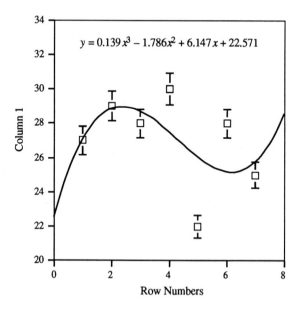

Fig. 12–9. CA-CRICKETGRAPH plot of the data illustrated in Fig. 12–8.

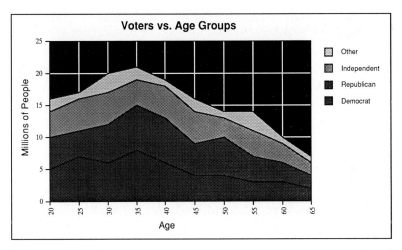

Fig. 12–10. Example of a demographic chart prepared with the aid of CA-CRICKETGRAPH III.

12.4 Spreadsheet Programs

Spreadsheet programs are most often associated with computational problems that arise in the world of business, but such programs also have their place in science. A spreadsheet program is essentially a tool for accomplishing a set of well-defined, routine transformations (mathematical, statistical, or graphic) with respect to data supplied initially in tabular form.[4] Organized around a format reminiscent of a matrix (in some cases a very large matrix!), data entries *(elements)* in particular rows and columns are caused to interact with one another in consistent, mathematically unambiguous ways. For example, in the program EXCEL (Microsoft), one of the leading spreadsheet applications for both the PC and MACINTOSH environments,[5] associating the command

    =SUM(B4:E4)

---

[4] As noted previously, many of the features of a typical spreadsheet program are also provided by the database program PANORAMA (Sec. 12.2.2).
[5] EXCEL's chief competitor in terms of the PC is 1-2-3 by Lotus, which only recently became available in a satisfactory MACINTOSH version.

with a particular *cell* (point of intersection of a column with a row) in the fourth row of a spreadsheet ensures that the cell in question will be filled automatically with the sum of the entries appearing in columns two through five of that same row. As our example suggests, *columns* in EXCEL (and most other programs of this type) are designated by letter, *rows* by number.

Spreadsheet software has only gradually come to be recognized as a potentially valuable adjunct to a scientific investigation, perhaps in consequence of unimaginative marketing on the part of software houses—or because the scientific community has been dismissed as "trivial" compared to the vast corporate market. Nevertheless, many scientists have found spreadsheet programs to be very useful, offering an elegant and rapid approach to the solution of a host of routine arithmetic problems. Thus, the simple act of entering primary experimental data into appropriately defined cells of a graphically displayed "worksheet" can evoke an extensive series of complex calculations, leading almost instantly to complete evaluation of the results. Subsidiary search functions can also be used to retrieve specific pieces of information, equivalent to working with a database manager, and powerful tools are typically available for carrying out statistical analyses or even transforming numerical data into professional-looking graphs. Microsoft's EXCEL also features an advanced programming language (masked as a "macro" function) that permits the experienced user to develop computational procedures far more complex than any envisioned by the developers.

Figure 12–11 illustrates with a simple example the type of calculation a chemist might wish to perform with the aid of an EXCEL spreadsheet.

12.5 HyperCard

Like other hardware manufacturers, Apple has long been conscious of the fact that consumers often base computer hardware decisions on the possibility of running unique software. This recognition certainly contributed to Apple's decision to encourage the development of HYPERCARD for the MACINTOSH, a remarkable system best described as a cross between a relational database (cf. Sec. 12.2.3), a programming language, and an integrated graphics environment. One reviewer has described HYPERCARD as a "multimedial information system"; Bill Atkinson, the principal developer, speaks of HYPERCARD

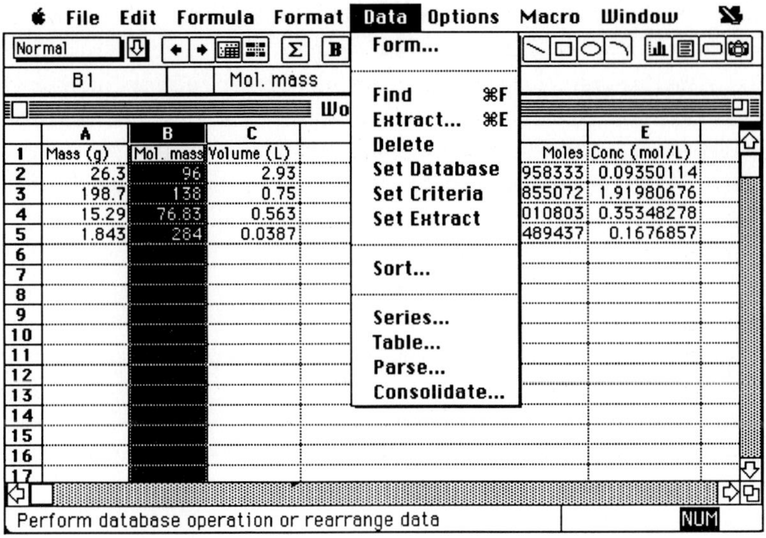

Fig. 12–11. Use of an EXCEL spreadsheet to carry out concentration calculations for a chemical system.

as a "screen information system"—in other words, a package pointed directly toward desktop presentations (cf. Sec. 1.6).[6]

The presence of the word "card" in HYPERCARD draws attention to the fact that the information in question is encapsulated in a format similar to a file card, but graphic images play an extremely important role in HYPER-CARD "file cards". The cards themselves are ordinarily grouped into "stacks". Thus, a particular stack might be designated for information from an address book, with each card bearing a single name, address, and phone number. As with other databases, subsets of cards can be displayed, printed, sorted in various ways, or subjected to a targeted search, and individual entries are always subject to change.[7]

What distinguishes a HYPERCARD data system most clearly from other databases, however, is the role assigned to "buttons": defined regions on the cards that are sensitive to mouse activation as a convenient way of initiating

---

[6] For more information on HYPERCARD in general see, for example, Beekman (1992), Goodman (1987), Michel (1989), or Sanders (1989).

[7] Though perhaps only by the *author*; see below.

some form of action. Buttons can be positioned almost anywhere, subject only to the whims of the stack designer, and their appearance is also a function of the designer's imagination (cf. Fig. 12–12). The primary purpose of HYPERCARD buttons is to establish links between various cards in various stacks. For example, if a button were to carry the label "Information", activating it with the mouse could be expected to lead to display of another card supplying details regarding its precursor.

It is the creative use of carefully conceived buttons that confers upon HYPERCARD its amazing versatility with respect to the structuring of information. A simple example is provided by the initial card in the original HYPERCARD master deck as prepared by Apple, referred to as the "Home

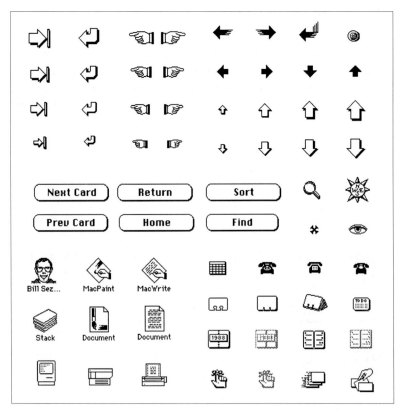

Fig. 12–12. A selection of "buttons" for use in conjunction with HYPERCARD.

Card" (Fig. 12–13). Each of the images on this card is in fact a button capable of transporting the user instantly to a different stack, one that contains additional information directly pertinent to that button.

A given HYPERCARD stack can be made user accessible at any of five levels of interaction *(authorization)*, the appropriate level being established through a dialog box.

- The lowest level of access, "browsing", is limited to viewing (or printing) the contents of the stack as it currently exists, without any provision for stack modification. Certain demonstration stacks supplied with the program itself operate at least initially at this level.

- A stack associated with the second authorization level ("typing") is subject to user modification, deletion, or amplification, but only with respect to the content of certain text fields.

At the first authorization level, user interaction is supported by only three menus, each with a very limited number of options. The purpose of this restriction—apart from security considerations—is to simplify program operation to such an extent that the inexperienced user will quickly come to feel comfortable with the medium. Thus, the usual MACINTOSH "File" and "Edit" menus are joined only by a "Go" menu that permits one to request

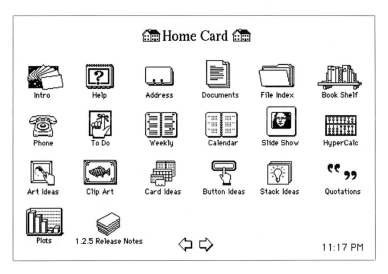

Fig. 12–13. The HYPERCARD "home stack".

access to a different card, or to institute an elementary search. The "typing" level is nearly as limited, adding only "Font" and "Style" menus.

- At the third level ("painting") the user acquires the right to alter cards with respect to their graphic content. This is accomplished through a "Tools" menu similar to that in the program MACPAINT (Sec. 11.2), providing a brush, a pencil, and an eraser, together with facilities for rotation, inversion, rescaling, etc. Graphics prepared with other software packages can also be imported and incorporated into HYPERCARD records.

The two highest levels of authorization permit a user to create and modify entire stacks, adding an important menu called "Objects".

- The fourth level ("authoring") facilitates creation of an unlimited number of buttons for linking together cards from various stacks.

- The final level, "scripting", opens the way to use of the HYPERCARD programming language, known as HYPERTALK.[8]

HYPERTALK is an "object-oriented" language: that is, it presupposes a data structure in which data and procedures can be combined to form new, larger entities called "objects". The objects themselves might constitute buttons, fields, individual cards, "backgrounds" (templates forming the basis for groups of cards, and specifying the layout of graphics, text fields, and buttons), or even complete stacks. Instead of preparing a single long program, a HYPERTALK "author" typically combines appropriate limited sets of command chains in such a way that they ultimately constitute an unambiguous description of a single larger object. Object-oriented programming requires that individual messages be directed toward specific objects. The object addressed may itself be in a position to execute the command in its entirety, or it may pass portions along to one or more additional objects. This type of programming is especially well suited to an *event-driven application*, which in the Macintosh environment represents the rule. Examples of an "event" in this sense would include pressing the mouse button or introducing a particular image or text fragment into a window of some type. Whether or not an object reacts to the event in question and, if so, how it reacts, is a function of the structure of the corresponding HYPERTALK *script* (program); thus,

---

[8]   Useful guides to the HYPERTALK language include Shafer (1991); Waite, Prata, and Jones (1989); Weiskamp and Shammas (1988); and especially Winkler (1990).

every text field, button, card, or stack is likely to be associated with its own HYPERTALK script.

The ease with which one is able to manipulate data constituting a HYPER-CARD stack tends to mask the incredible versatility inherent in the HYPERTALK programming language. The language is based on everyday speech, but it allows one to write programs of remarkable subtlety. In particular, HYPER-CARD buttons can be programmed to do far more than simply transfer control to a different card, such as initiating searches subject to multiple criteria, performing wide-ranging calculations, and providing for unusual forms of output.

HYPERTALK incorporates a wealth of commands and functions, including standard sequences of the "IF ... THEN ... ELSE ..." and "REPEAT ... UNTIL ..." type. The GOTO command in HYPERTALK serves a somewhat unusual purpose: it transfers control not to a different part of the program, but rather to a different card. HYPERTALK also includes provisions for creating or deleting an unlimited number of stacks, cards, buttons, or text fields, as well as special commands for accessing (and altering) external data files. Procedures developed originally in other programming languages (e.g., Pascal) can easily be linked to HYPERTALK commands and functions. HYPERCARD cards can even be programmed to incorporate dynamic images, prepared separately with the aid of a program like ADOBE PREMIER (Adobe Systems).

When it was first introduced, HYPERCARD revealed an entirely new dimension in the relationship between data management and data presentation. It has been found to be particularly valuable in educational applications (*teachware* or *courseware*),[9] especially with respect to complex technical disciplines. Thus, a HYPERCARD stack effectively endows a computer with the ability to "take the learner by the hand" and guide his or her progress—gently, at an individualized pace—until a particular task or concept has been mastered. At the same time, the user retains full control over the actual learning situation, free to move at will from one point in a lesson to another.

HYPERCARD makes it relatively easy for one to create highly sophisticated data resources, didactic presentations, or illustrated catalogues. Indeed, several corporations have experimented with HYPERCARD-based consumer

---

[9] HYPERCARD has become a source of countless commercial applications packages (often referred to as "stackware").

catalogues, complete with scanned images of the products on offer as well as convenient electronic order forms—which can in turn be linked to inventory-control records—in an attempt to take optimum advantage of this striking and effective format, which also lends itself to convenient and frequent editing and updating.

12.6 Special-Purpose Software

We conclude with a very brief survey of other types of software likely to be of broad interest to scientists and engineers active in desktop publishing. Observations here will be limited to a mention of various *categories* of software and their purpose, together with name identifications of one or two examples of corresponding resources available for use with the MACINTOSH operating system.

● Bibliographic software

Programs in this category are really special-purpose flat-file databases that have been adapted to the unique requirements of bibliographic citation. A "citation manager" works in conjunction with a word processor in such a way that direct links are established between a master database and the citation numbers entered into a manuscript, permitting rapid assembly of a complete reference list. Database templates are usually provided for each of several different document categories (books, journal articles, patents, etc.), with each template containing appropriate data fields for all the bibliographic elements required to characterize that particular type of document. Moreover, extensive provisions are made for outputting the data in various ways in an effort to meet the often contradictory demands of different publishers (for more information on the complex subject of bibliographic citation see Ebel, Bliefert, and Russey, 1987, Chap. 11). The leading program of this type for the MACINTOSH is ENDNOTE PLUS (Niles & Associates).

● Resource Expansion Tools

Every desktop publisher sooner or later encounters hardware limitations in the form of insufficient disk storage space and/or insufficient memory. The long-term solution is of course an investment in new facilities, but software solutions can provide a temporary respite. Thus, there now exist a number of excellent file-compression utilities capable of recovering a considerable amount of disk-storage space without also imposing an unreasonable penalty in terms of lost time. Both the DISKDOUBLER (Salient Software) and STUFFIT (Alladin Systems) utilities for the MACINTOSH work very efficiently in this respect, typically recovering 50% or more of the space otherwise required for data storage. The savings can be especially dramatic with bitmap graphics files (cf. Sec. 5.5). A relatively new program called RAM DOUBLER (Connectix) is able to work similar magic with respect to the computer's memory, inducing a MACINTOSH to "think" it contains twice as many memory chips as are actually present—with essentially no noticeable effect on performance, and at an almost unbelievably low price (cf. Sec. 5.6).

● Macro Utilities

Many desktop publishing operations are so routine that one often wishes it were possible personally to program the computer to carry out "nuisance tasks" upon issuance of a simple keystroke command. This can in fact be done, and quite easily—with a "macro" utility like QUICKEYS (CE Software) or TEMPO (Affinity Microsystems).

● "Productivity Aids"

Apart from serving as a work station, the computer can also play a useful role in conferring additional order upon one's personal and professional life. Programs like DATEBOOK and TOUCHBASE (both from Aldus) or INCONTROL (Attain Corp.) encourage an organized approach to the burdens of scheduling and project planning.

* * *

The temptation was great to conclude with suggestions for recreational software useful on those not infrequent occasions when a particular writing or page layout assignment seems destined to culminate only in frustration, but we ultimately elected instead to leave this aspect of electronic writing entirely in the hands of the resourceful reader.

Appendixes

Appendix A
Cited Literature

Adobe Systems, Inc. (Ed.) (1985): *PostScript Language—Tutorial and Cookbook*. Reading, Mass.: Addison-Wesley. 243 p.

Adobe Systems, Inc. (Ed.) (1988): *PostScript Language—Program Design*. Reading, Mass.: Addison-Wesley. 224 p.

Adobe Systems, Inc. (Ed.) (1990): *PostScript Language—Reference Manual*. 2nd ed. Reading, Mass.: Addison-Wesley. 600 p.

Armstrong, C., Large, J. (Eds.) (1992): *Manual of Online Search Strategies*. 2nd ed. New York: G. K. Hall. 699 p.

Baeseler, F., Heck, B. (1987): *Desktop Publishing*. Hamburg: McGraw-Hill. 266 p.

Barolini, H. (1992): *Aldus and His Dream Book*. New York: Italica Press. 222 p.

Baumann, H. D., Klein, M. (1992): *Desktop Publishing: Typografie, Layout Seiten gestalten am PC für Profis und Einsteiger*. Niedernhausen, Germany: Falken. 320 p.

Beach, M., Russon, K. (1989): *Papers for Printing: How to Choose the Right Paper at the Right Price for Any Printing Job*. Manzanita, Ore.: Coast to Coast. 160 p.

Bechtolsheim, S. (1992): *TeX in Practice*. 4 Vols. (I: *Basics*, 384 p.; II: *Paragraphs, Math, & Fonts*, 368 p.; III: *Tokens, Macros*, 544 p.; IV: *Output Routines, Tables*, 576 p.) (D. Rogers, Ed.). New York: Springer.

Beekman, G. (1992): *HyperCard 2 in a Hurry*. Belmont, Cal.: Wadsworth Pub. Co. 340 p.

Beyer, W. H. (Ed.) (1987): *CRC Handbook of Mathematical Sciences*. 6th ed. Boca Raton, Fla.: CRC Press. 860 p.

Blatner D., Stimely, K. (1991): *The QuarkXPress Book*. 2nd ed. (S. Roth, Ed.). Berkeley: Peachpit Press. 640 p.

Blatner, D. (1991): *Desktop Publisher's Survival Kit*. Berkeley: Peachpit Press. 184 p.

Blatner, D., Roth S. (1993): *Real World Scanning & Halftones.* Berkeley: Peachpit Press. 276 p.

Bosshard, H. R. (1980): *Technische Grundlagen zur Satzherstellung.* Bern: Verlag des Bildungsverbandes Schweizerischer Typografen BST. 296 p.

Bottle, R. T., Rowland, J. F. B. (Eds.) (1992): *Information Sources in Chemistry.* 4th ed. London: Bowker-Saur. 341 p.

Bove, T., Rhodes, C., Thomas, W. (1986): *The Art of Desktop Publishing— Using Personal Computers To Publish It Yourself.* Toronto: Bantam. 296 p.

Bradley, J. C. (1992): *Desktop Publishing.* Dubuque: Wm. C. Brown. 496 p.

Braswell, F. M. (1989): *Inside PostScript.* Berkeley: Peachpit Press. 328 p.

Braun, E. (1994): *The Internet Directory.* New York: Fawcett Columbine, 704 p.

Breuer, M. (1988): *Das PageMaker-Praxis-Buch – Professionelles Publizieren mit dem PC.* Haar, Germany: Markt & Technik. 410 p.

Bringhurst, R. (1992): *The Elements of Typographic Style.* Point Roberts, Wash.: Hartley & Marks. 256 p.

Brown, A. (1989): *In Print—Text and Type in the Age of Desktop Publishing.* New York: Watson-Guptill. 192 p.

Bryan, M. (1988): *SGML: An Author's Guide to the Standard Generalized Markup Language.* Reading, Mass.: Addison-Wesley. 364 p.

Busch, D. D. (1991): *The Complete Scanner Handbook for Desktop Publishing, Macintosh Edition.* Homewood, Ill.: Business One Irwin. 386 p.

Busche, D., Glenn, B. (1991): *Desktop Design Workbook.* New York: Prentice-Hall. 320 p.

Card, M. (1990): *Word Processor to Printed Page—A Guide to Interfacing Word Processors and Phototypesetters.* 2nd ed. New York: Van Nostrand Reinhold. 181 p.

Cavuoto, J., Beck, S. (1992): *The GATF Guide to Desktop Publishing.* Pittsburgh: Graphic Arts Technical Foundation. 200 p.

Chicago Guide to Preparing Electronic Manuscripts for Authors and Publishers (1987). Chicago: The University of Chicago Press. 143 p.

Cole, M., Odenwald, S. (1989): *Desktop Presentations.* New York: AMACOM. 250 p.

Cookman, B. (1993): *Desktop Design: Getting the Professional Look.* New York: Van Nostrand Reinhold. 120 p.

Craig, J. (1992): *Designing with Type—A Basic Course in Typography.* 3rd rev. ed. New York: Watson-Guptill. 176 p.

Day, R. A. (1983): *How to Write and Publish Scientific Papers.* 2nd ed. Philadelphia: ISI Press. 182 p.

Dern, D. (1994): *The Internet Guide for New Users.* New York: McGraw-Hill. 570 p.

Désarménien, J. (1986): *TeX for Scientific Documentation (LCNS 236).* Berlin: Springer. 204 p.

DIN Deutsches Institut für Normung e.V. (Ed.) (1986): *DIN-Fachbericht 4: Graphische Symbole nach DIN 30 600, Teil 1: Bildzeichen, Übersicht.* 5th ed. Berlin: Beuth-Verlag. 108 p.

Ebel, H. F., Bliefert, C., Russey, W. E. (1987): *The Art of Scientific Writing —From Scientific Reports to Professional Publications in Chemistry and Related Fields.* New York: VCH Publishers. 494 p.

Faulmann, C. (1990): *Das Buch der Schrift – Enthaltend die Schriftzeichen und Alphabete aller Zeiten und aller Völker des Erdkreises.* Reprint of the 1880 edition. Frankfurt: Eichborn Verlag. 286 p.

Fenton, E. (1991): *The Macintosh Font Book.* 2nd ed. Berkeley: Peachpit Press. 352 p.

Gerstner, K. (1985): *Kompendium für Alphabeten – eine Systematik der Schrift.* Teufen, Germany: Arthur Niggli.

Gilster, P. (1993): *The Internet Navigator.* New York: Wiley. 470 p.

Glover, G. (1989): *Image Scanning for Desktop Publishers.* Blue Ridge Summit, Pa.: TAB Books. 400 p.

Goldfarb, C. F. (1990): *The SGML Handbook.* Oxford: Oxford University Press. 550 p.

Goodman, D. (1987): *The Complete HyperCard Handbook.* New York: Bantam Books. 720 p.

Gosney, M., Dayton, L. (1990): *The Verbum Book of Electronic Page Design.* Redwood City, Cal.: M&T Books. 212 p.

Gosney, M., Dayton, L., Chang P. (1991): *The Verbum Book of Scanned Imagery.* Redwood City, Cal.: M&T Books. 211 p.

Grosvenor, J., Morrison, K., Pim, A. (1992): *The Postscript™ Font Handbook—A Directory of Type 1 Fonts.* Rev. ed. Wokingham, U. K.: Addison-Wesley Publishers Ltd. 426 p.

Günder, G. (1988): *Desktop Design.* Düsseldorf: Publishing Partner. 120 p.

Gurari, E. (1993): *TeX and LaTeX: Drawing and Literate Programming.* New York: McGraw-Hill. 250 p.

Hahn, H., Stout, R. (1994): *The Internet Complete Reference.* Berkeley: Osborne McGraw-Hill. 818 p.

Hartley, R. J. et al. (1990): *Online Searching, Principles and Practice.* New York: Bowker-Saur. 387 p.

Heck, A. (1992): *Desktop Publishing in Astronomy and Space Sciences.* River Edge, N.J.: World Scientific Publishers. 240 p.

Hewson, D. (1988): *Introduction to Desktop Publishing—A Guide to Buying and Using a Desktop Publishing System.* San Francisco: Chronicle Books. 112 p.

Hochuli, J. (1987): *Das Detail in der Typografie – Buchstabe, Buchstabenabstand, Wort, Wortabstand, Zeile, Zeilenabstand, Kolumne.* Wilmington, Mass.: Compugraphic. 48 p.

Holzgang, D. A. (1987): *Understanding PostScript Programming.* San Francisco: Sybex. 472 p.

IUPAC International Union of Pure and Applied Chemistry (Ed.) (1988): *Quantities, Units, and Symbols in Physical Chemistry* (1988). Oxford: Blackwell Scientific Publications. 134 p.

Iyanaga, S., Kawada, Y. (Eds.) (1977): *Encyclopedic Dictionary of Mathematics.* Cambridge, Mass.: MIT Press. 1750 p.

Jean, G. (1992): *Writing—The Story of Alphabets and Scripts.* New York: Harry Abrams. 208p.

Jones, G. (1987): *Desktop Publishing Companion.* Wilmslow, U. K.: Sigma Press. 202 p.

Kist, J. (1988): *Elektronisches Publizieren – Übersicht, Grundlagen, Konzepte.* Stuttgart: Raabe. 180 p.

Kleper, J. L (1990): *The Illustrated Handbook of Desktop Publishing.* 2nd ed. Blue Ridge Summit, Pa.: Windcrest Books. 927 p.

Knuth, D. E. (1987): *The TeXbook.* Vol. 2 of the series *Computers and Typesetting.* 11th ed. Reading, Mass.: Addison-Wesley and the American Mathematical Society. 484 p.

Korger, H. (1977): *Schrift und Schreiben.* 3rd ed. Leipzig: VEB Fachbuchverlag. 264 p.

Kvern, O., Roth, S. (1992): *PageMaker Tips & Tricks: Industrial Strength Techniques.* New York: Bantam.

Kvern, O., Roth, S. (1993): *PageMaker Tips & Tricks: Secrets of the PM Masters*. Windows edition. Berkeley: Peachpit Press. 320 p.

Lane, E., Summerhill, C. (1992): *Internet Primer for Information Professionals: A Basic Guide to Internet Networking Technology*. Westport, Conn.: Meckler. 175 p.

Lawson, A. (1990): *Anatomy of a Typeface*. Boston: David R. Godine. 432 p.

Lee, M. (1980): *Bookmaking: The Illustrated Guide to Design/Production/Editing*. 2nd ed. New York: R. R. Bowker. 486p.

Lerner, R. G., Trigg, G. L. (Eds.) (1991): *Encyclopedia of Physics*. 2nd ed. New York: VCH Publishers. 1408 p.

LeVitus, B., Ihnatko, A. (1992): *Dr. Macintosh's Guide to the Online Universe*. Reading, Mass.: Addison-Wesley. 362 p.

Lieberman, J. B. (1967): *Types of Typefaces—and How to Recognize Them*. New York: Sterling Publishing Co. 132 p.

Luidl, P. (1984): *Typographie – Herkunft, Aufbau, Anwendung*. Hannover: Schlütersche Verlagsanstalt und Druckerei. 146 p.

Luidl, P. (1988): *Desktop-Knigge – Setzerwissen für Desktop-Publisher*. Munich: te-wi. 198 p.

Marshall, G. R. (1990): *The Manager's Guide to Desktop Electronic Publishing*. Englewood Cliffs, N. J.: Prentice-Hall. 267 p.

McClelland, D., Danuloff, C. (1987): *Desktop Publishing Type & Graphics – A Comprehensive Handbook*. Boston: Harcourt Brace Jovanovich. 264 p.

McLean, R. (1980): *The Thames and Hudson Manual of Typography*. London: Thames and Hudson. 216 p.

McNamara, J., Romkey, J. (1993): *Local Area Networks*. 2nd ed. Englewood Cliffs, N. J.: Prentice-Hall. 225 p.

McNeil, D., Freiberger P. (1993): *Fuzzy Logic: The Discovery of a Revolutionary Computer Technology—and How It Is Changing Our World*. New York: Simon & Schuster. 320 p.

Meehan, T., et al. (1992): *QuarkXPress 3.1 by Example*. San Mateo, Cal.: M&T Books. 338 p.

Menousos, S., Tilden, S. W. (1990): *The Professional Look: The Complete Guide to Desktop Publishing*. Santa Barbara, Cal.: Venture Perspective Press. 275 p.

Meyer, D. E., Warr, W. A., Love, R. A. (Eds.) (1988): *Chemical Structure Software for Personal Computers*. ACS Professional Reference Book. Washington, D. C.: American Chemical Society. 108 p.

Michel, S. (1988): *HyperCard, the Complete Reference*. Berkeley, Cal.: Osborne McGraw-Hill. 730 p.

Motorola Codex (1993): *The Basics Book of Information Networking*. Reading, Mass.: Addison-Wesley. 176 p.

Nance, B. (1993): *Introduction to Networking*. Carmel, Ind.: Que. 512 p.

National Research Council, Computer Science and Telecommunications Board (1991): *Intellectual Property Issues in Software*. Washington, D. C.: National Academy Press. 112 p.

Parker, C. C., Turley, R. V.: *Information Sources in Science and Technology, a Practical Guide to Traditional and Online Use*. 2nd ed. London: Butterworths. 328 p.

Parker, S. P. (Ed.) (1989): *McGraw-Hill Dictionary of Scientific and Technical Terms*. 4th ed. New York: McGraw-Hill. 2140 p.

Perfect, C., Rookledge, G. (1991): *Rookledge's International Typefinder: The Essential Handbook of Typeface Recognition and Selection*. Rev. ed. Mount Kisko, N. Y.: Moyer Bell. 284 p.

Peters, J. (1988): *Desktop Publishing. Was bringt's wirklich: – Analysen, Erfahrungen, Umfeld; Hardware, Software, Brainware*. Wiesbaden: Gabler. 140 p.

Potkin, N., Hansen, H., Schneider, D. (eds.) (1992): *The 1993 BMUG Shareware Disk Catalog*. Reading, Mass.: Addison-Wesley. 686 p.

Rüegg, R (1972): *Typografische Grundlagen – Handbuch für Technik und Gestalten*. Zurich: ABC Verlag. 220 p.

Sanders, W. B. (1989): *HyperCard Made Easy*. Glenview, Ill.: Scott, Foresman. 408 p.

Schulze, H. H. (1988): *Das Rororo Computer Lexikon*. Reinbek, Germany: Rowohlt. 568 p.

Shafer, D. (1991): *The Complete Book of HyperTalk 2*. Rev. ed. Reading, Mass.: Addison-Wesley. 444 p.

Simrin, S. (1988): *The Waite Group's MS-DOS Bible*. 2nd ed. Indianapolis: Howard W. Sams & Co. 524 p.

Sitarz, D. (1989): *Desktop Publisher's Legal Handbook: A Comprehensive Guide to Computer Publishing Law*. Carbondale, Ill.: Nova Publishing Co. 240 p.

Smith, J. M. (1986a): *The Standardized Generalized Markup Language (SGML): Guidelines for Editors and Publishers*. Boston Spa, U. K.: The British Library.

Smith, J. M. (1986b): *The Standardized Generalized Markup Language (SGML) and Related Issues.* British National Bibliography Research Fund Report 22. Boston Spa, U. K.: The British Library.

Smith, J. M. (1987): *The Standardized Generalized Markup Language (SGML): Guidelines for Authors.* British National Bibliography Research Fund Report 27. Boston Spa, U. K.: The British Library.

Smith, R. (1990): *Learning PostScript: A Visual Approach.* Berkeley: Peachpit Press. 350 p.

Snow, W. (1992): *TeX for the Beginner.* Reading, Mass.: Addison-Wesley. 320 p.

Spencer, H. (1969): *The Visible Word.* 2nd rev. ed. New York: Hastings House. 107 p.

Spiekermann, E. (1986): *Ursache & Wirkung: ein typografischer Roman.* Erlangen, Germany: Context. 128 p.

Spivak, M. (1990): *Joy of TeX.* Rev. ed. Providence: American Mathematical Society. 309 p.

Stallings, W. (1993): *Data and Computer Communications.* 4th ed. New York: Macmillan. 864 p.

Straka, R., Dickschus, A., Leyhausen, M., Kneisch, K.-D. (1987): *Electronic Publishing – Personalcomputer in Druckerei, Werbung und Verlag.* Haar, Germany: Markt & Technik. 328 p.

Strunk, W., Jr., White, E. B. (1979): *The Elements of Style.* 3rd ed. New York: MacMillan. 92 p.

Swanson, E. (1988): *Mathematics into Type: Copyediting and Proofreading of Mathematics for Editorial Assistants and Authors.* 2nd. ed., rev. Providence: American Mathematical Society. 98 p.

The Chicago Manual of Style (1982). 13th ed. Chicago: University of Chicago Press. 738 p.

The Cobb Group Staff (1993): *The Word 5.1 Companion—Mac.* Redmont, Wash.: Microsoft Press. 687 p.

van Herwijnen, E. (1990): *Practical SGML.* Dordrecht: Kluwer Academic Publishers. 307 p.

Vollenweider, P. (1988): *PostScript – Konzeption, Anwendung, Mischen von Text und Grafik.* Munich: Carl Hanser. 178 p.

Waite, M., Prata, S., Jones, T. (1989): *The Waite Group's HyperTalk Bible.* Indianapolis: Hayden Books. 692 p.

Warr, W. A. (Ed.) (1987): *Graphics for Chemical Structures—Integration with Text and Data*. ACS Symposium Series No. 341. Washington, D. C.: American Chemical Society. 176 p.

Weiskamp, K., Shammas, N. (1988): *Mastering HyperTalk*. New York: Wiley. 506 p.

White, J. V. (1988): *Graphic Design for the Electronic Age—The Manual for Traditional and Desktop Publishing*. New York: Watson-Guptill (a Xerox Press Book). 212 p.

Winkler, D., Kamins, S. (1990). *Hypertalk 2.0, the Book*. New York: Bantam Books. 958 p.

Yasui, H. (1989): *Desktop Publishing: Technology and Design*. Eden Prairie, Minn.: Paradigm Publishing International. 256 p.

Zech, R. (1985): *Das Programmiersprache FORTH*. Munich: Franzis. 334 p.

Ziegfeld, R., Tarp, J. (1989): *Desktop Publishing for the Writer: Designing, Writing, and Developing*. Piscataway, N. J.: IEEE Computer Society Press. 362 p.

Appendix B
Addresses

Abacus Concepts, Inc., 1984 Bonita Ave., Berkeley, CA 94704; (800) 666-7828, (510) 540-1949

ACIUS, Inc., 10351 Bubb Rd., Cupertino, CA 95014; (408) 252-4444

Adobe Systems, 1585 Charleston Rd., P.O. Box 7900, Mountain View, CA 94039; (800) 833-6687, (415) 961-4400

Advanced Software, Inc., 1095 E. Duane Ave., Suite 103, Sunnyvale, CA 94086; (800) 346-5392, (408) 733-0745

Affinity Microsystems Ltd., 1050 Walnut St., Suite 425, Boulder, CO 80302; (800) 367-6771, (303) 442-4840

Agfa Compugraphic—Agfa Division, Miles Inc. (Type Products), 200 Ballardvale St., Wilmington, MA 01887; (800) 227-2780, (508)658-5600

Aladdin Systems, Inc., 165 Westridge Dr., Watsonville, CA 95076; (480) 761-6200

Aldus Corp., 411 First Ave. S., Seattle, WA 98104; (800) 333-2538; (206) 628-2320

Alki Software Corp, 219 First Ave. N., Suite 410, Seattle, WA 98109; (800) 669-9673, (206) 286-2600

Altsys Corp., 269 W. Renner Rd., Richardson, TX 75080; (214) 680-2060

Apple Computer, Inc., 20525 Mariani Ave., Cupertino, CA 95014; (800) 776-2333, (408) 996-1010

Ares Software, P.O Box 4667, 561 Pilgrim Dr., Suite D, Foster City, CA 94404; (800) 783-2737, (415) 578-9090

Attain Corp., 48 Grove St., Somerville, MA 02144; (617) 776-1110

Autodesk, Inc., 2320 Marinship Way, Sausalito, CA 94965; (415) 332-2344; Autodesk Retail Products, 11911 N. Creek Parkway S., Bothell, WA 98011; (800) 228-3601, (206) 487-2233

BMUG (Berkeley Macintosh Users Group), 1442A Walnut Street #62, Berkeley, CA 94709; (510) 849-9026

Blue Sky Research, 534 S.W. Third Ave., Portland, OR 97204; (800) 622-8398, (503) 222-9571

Borland International, Inc., 1800 Green Hills Rd., Scotts Valley, CA 95066; (800) 331-0877, (408) 438-5300

Brooks/Cole Publishing Co., 511 Forest Lodge Rd., Pacific Grove, CA 93950; (408) 373-0728

Caere Corp., 100 Cooper Court, Los Gatos, CA 95030; (800) 535-7226, (408) 395-7000

Calera Recognition Systems, 475 Potrero Ave., Sunnyvale, CA 94086; (800) 422-5372, (408) 720-8300

Cambridge Scientific Computing, 875 Massachusetts Avenue, Sixth Floor, Cambridge, MA 02139; (617) 491-6862

Casady & Greene, Inc., 22734 Portola Dr., Salinas, CA 93908; (800) 359-4920, (408) 484-9228

CE Software, 1801 Industrial Cir., P.O Box 65580, West Des Moines, IA 50265; (800) 523-7638, (515) 224-1995

Central Point Software, Inc., 15220 N.W. Greenbrier Parkway, Suite 200, Beaverton, Or 97006; (800) 445-4208, (503) 690-8088

Claris Corp., 5201 Patrick Henry Dr., Santa Clara, CA 95052; (408) 727-8227

Computer Associates International, Inc., One Computer Associates Plaza, Islania, NY 11788; (800) 531-5236, (516) 342-5224

Computervision (Personal CAD/CAM Business Unit), 100 Crosby Drive, Bedford, MA 01730; (800) 488-7228, (617) 275-1800

Connectix Corp., 2600 Campus Dr., San Mateo, CA 94403; (800) 950-5880, (415) 571-5100

Corel Corporation, 1600 Carling Ave., Ottawa, Ontario, K1Z 8R7 (Canada); (613) 728-3733

Dantz Development Corp., 1400 Shattuck Ave., Suite 1, Berkeley, CA 94709; (510) 849-0293

Datalogics, Inc., 441 W. Huron St., Chicago, IL 60610; (312) 266-4444

Datawatch, PO Box 51489, Durham, NC 27717; (919) 490-1277

Delta Point Inc., 2 Harris Court, Suite B-1, Monterey, CA 93940; (800) 367-4334, (408) 648-4000

Deneba Software, 7400 S.W. 87th Age, Miami, FL 33173; (305) 596-5644

Design Science Inc., 4028 Broadway, Long Beach, CA 90803; (800) 827-0685, (310) 433-0685

Electronic Arts, 1450 Fashion Island Blvd., San Mateo, CA 94404; (800) 245-4525, (415) 571-7171

Fifth Generation Systems, 10049 N. Reiger Rd., Baton Rouge, LA 70809; (800) 873-4384, (504) 291-7221

Fractal Design Corp., 335 Spreckels Dr., Suite F, Aptos, CA 95003; (408) 688-8800

Frame Technology Corp., 1010 Rincon Circle, San Jose, CA 95131; (800) 843-7263, (408) 433-3311

FreeSoft Co., 105 McKinley Rd., Beaver Falls, PA 15010 (412) 846-2700

Golden Triangle Computers, Inc., 4849 Ronson Court, San Diego, CA 92111; (800) 326-1858, (619) 279-2100

GraphPad Software [**InStat**]; (619) 457-3909

Graphsoft, Inc., 8370 Court Ave., Suite 202, Ellicott City, MD 21043; (410) 461-9488

Hell-Linotype, 425 Oser Ave., Hauppauge, NY 11788; (800) 633-1900, (516) 434-2000

Hewlett-Packard (direct Marketing), P.O. Box 58059, MS#511L-SJ, Santa Clara, CA 95051; (800) 752-0900

Icon Technologies, 317 Varsity Park #202, Charlottesville, VA 22903; (804) 977-1551

Imagen Corp., 2650-T San Tomas Expressway, Santa Clara, CA 95052; (408) 986-9400

Innovative Data Design, Inc., 135D Mason Circle, Concord, CA 94520 (510) 680-6818

Insignia Solutions, 526 Clyde Ave., Mountain View, CA 94043; (415) 694-7600

Interactive Solutions, Inc., 53-T W. Fort Lee Rd., Bogota, NJ 07603; (201) 488-3708

Kensington Microware, 2855 Campus Dr., San Mateo, CA 94403; (800) 535-4242, (415) 572-2700

Letraset USA, Inc., 40 Eisenhower Dr., Paramus, NJ 07653; (800) 343-8973, (201) 845-6100

Light Source, 17 E. Sir Francis Drake Blvd., Suite 100, Larkspur, CA 94939; (800) 231-7226, (415) 461-8000

Logitech, Inc., 6505 Kaiser Dr., Fremont, CA 94555; (800) 231-7717, (510) 795-8500

Lotus Development Corp., 55 Cambridge Parkway, Cambridge, MA 02142; (800) 688-8320, (617) 577-8500

MacVonk, Inc., 940 Sixth Ave. S.W., Suite 1100, Calgary, Alberta T2P3T1 (Canada); (403) 232-6545

MacWarehouse, 1720 Oak Street, P.O. Box 3013, Lakewood, NJ 08701; (800) 255-6227, (908) 370-4779

Manhattan Graphics Corp., 250 E. Hartsdale Ave., Suite 23, Hartsdale, NY 10530; (800) 572-6533, (914) 725-2048

MathSoft, Inc., 201 Broadway, Cambridge, MA 02139; (800) 628-4223, (617) 577-1017

Micrografx, P.O Box 850187, Richardson, TX 75085; (800) 733-3729

Microsoft Corp., One Microsoft Way, Redmond, WA 98052; (800) 426-9400, (206) 882 8080

Microspot USA, Inc., 20421 Stevens Creek, Cupertino, CA 95014; (408) 253-2000

Minitab, Inc., 3081 Enterprise Drive, State College, PA 16801; (800) 448-3555, (814) 238-3280

Niles & Associates, Inc., 2000 Hearst St., Suite 200, Berkeley, CA 94709; (510) 649-8176

Nisus Software, Inc., P.O Box 1300, 107 S. Cedros, Solana Beach, CA 92075; (800) 922-2993, (619) 481-1477

Olduvai Corp., 7520 Red Rd., Suite A, South Miami, FL 33143 (800) 822-0772, (305) 665-4665

ON Technology, Inc., 155 Second St., Cambridge, MA 02141; (617) 876-0900

Paper Direct, 205 Chubb Ave., Lyndhurst, NJ 07071; (800) 272-7377, (201) 507-5488

Pergamon Press, Fairview Park, Elmsford, NY 10523; (914) 592-7700

PrairieSoft, Inc., P.O. Box 65820, West Des Moines, IA 50265; (515) 225-3720

Prescience Corp., 939 Howard St., San Francisco, CA 94103; (800) 827-6284, (415) 543-2252

ProVue Development Development, Inc., 15180 Transistor Lane, Huntington Beach, CA 92649; (714) 982-8199

Quark, Inc., 1800 Grant St., Denver, CO 80203; (800) 788-7835, (303) 894-8888

Ray Dream, Inc., 1804 N. Shoreline Blvd., Mountain View, CA 94043; (415) 960-0765

Salient Software, Inc., 124 University Ave., Suite 300, Palo Alto, CA 94301; (415) 321-5375

SoftQuad Inc., 56 Aberfoyle Crescent, Suite 810, Toronto, Canada M8X 2W4; (416) 239-7105

Softshell International, 715 Horizon Drive, Suite 390, Grand Junction, CO 81506; (303) 242-7502

Softways, 5066 El Roble, San Jose, CA 95118; (408) 978-9167

Specular International, P.O. Box 888, Amherst, MA 01004; (413) 549-7600

SPSS, Inc., 444 N. Michigan Ave., Chicago, IL 60611; (800) 543-6609, (312) 329-2400

SuperMac Technology, 485 Potrero Ave., Sunnyval, CA 94086; (800) 334-3005, (408) 245-2202

Symantec Corp., 10201 Torre Ave., Cupertino, CA 95014; (800) 441-7243, (408)-253-9600

ThunderWare, Inc., 21 Orinda Way, Orinda, CA 94563; (800) 445-1166, (510) 254-6581

Timeworks, Inc., 625 Academy Dr., Northbrook, IL 60062; (800) 535-9497, (708) 559-1300

Varityper, 11 Mt. Pleasant Ave., East Hanover, NJ 07936 (800) 631-8134, (201) 887-8000

Ventura Software, 15175 Innovation Dr., San Diego, CA 92128; (800) 822-8221, (619) 673-0172

Wolfram Research, Inc., 100 Trade Center Drive, Champaign, IL 61820; (800) 441-6284, (217) 398-0700

WordPerfect Corp., 1555 N. Technology Way, Orem, UT 84057; (800) 451-5151, (801) 225-5000

WordStar International, Inc., 201 Alameda del Prado, Novato, CA 94949; (800) 523-3520, (415) 382-8000

Working Software Inc., P.O. Box 1844, Santa Cruz, CA 95061; (408) 423-5696

Xerox Imaging Systems, Inc., 9 Centennial Dr., Peabody, MA 01960; (800) 248-6550, (508) 977-2000

XTree Co., 4330 Santa Fe Rd., San Luis Obispo, CA 93401; (805) 541-0604

Zedcor, 4500 E. Speedway, Suite 22, Tucson, AZ 85712; (800) 482-4567, (602) 881-8101

Z Soft Corp., 450 Franklin Rd., Suite 100, Marietta, GA 30067; (404) 428-0008

Appendix C
Remarks Concerning the Production of This Book

All of the work of preparing this book—writing and editing the text, creating the illustrations, and completing the layout—was accomplished with computers in the Apple MACINTOSH family (Classic II, SE/30, and Mac II).

Text was first prepared with MICROSOFT WORD (Versions 4.01 and 5.1) and then imported directly into PAGEMAKER 4.01 files.

Type in the body of the text is 12-point Times Roman with 4 points of interline spacing (text breadth: 132 mm). Footnotes, figure captions, and tables are set in 11-point Times with 2 points of interline spacing.

Pages were reduced during production to 80% of their original size. Three pages (146, 148, and 150) were reproduced from 300-dpi originals prepared with an Apple LASERWRITER PLUS; all other pages were prepared directly as films with a Linotronic 200 imagesetter (1693-dpi resolution).

Most of the illustrations—especially the screen shots—were reduced photographically. Only figures 2–3, 2–5, 2–9, 2–10, 2–13, 2–15, and 11–9 were prepared directly with the Linotronic 200.

Index

Ebel, H.F. /Bliefert, C. /Russey, W.E.

The Art of
Scientific Writing

From Student Reports to Professional
Publications in Chemistry and Related
Fields

1987. XIX, 493 pages with 27 figures and 16
tables. Hardcover. DM 105.-. ISBN 3-527-26469-8

First of all, here we have a highly readable text on
scientific communication - with all its ramifications
from writing the most effective text to getting it
published. Next, the handbook will also serve as a
ready-reference handbook for every emergency an
author may encounter (an important galley correction
has not been made, a Note Added in Proof must be
inserted, a copyright contract must be studied).
Finally, the volume is a complete and lucid description
of all methods and processes used in writing and
printing - to be looked up if an unfamiliar technical
term is encountered. It is this comprehensiveness that
gives the book its unique value. Never before has so
much information on scientific writing been assembled
in a single monograph.
Every situation is provided for - from the student's
first struggle with his report to his thesis to his journal
publications (where the author will relish the detailed
chapters on standardized terminology) to the manuscript of a book.
The author's tools are examined, with particular
attention to the latest developments in the generation
of text and graphics; there is generous advice available
on tables and references.
From the lonely labour at the author's desk the
discourse leads on to interaction with the author's
peers (editors, reviewers) and with publishers.

Date of information:
October 1994

VCH, P.O. Box 10 11 61,
D-69451 Weinheim,
Telefax (0) 62 01 - 60 61 84

Schoenfeld, R.

The Chemist's English

Third, revised edition
with Say It in English, Please!

1990. XII, 195 pages with 9 figures. Hardcover.
DM 48.-. ISBN 3-527-28003-0

Comments on this book:

"Chemists owe it to themselves to read this book,
and a high proportion of those who do are likely to
want a copy to keep all to themselves ..."
Journal of the American Chemical Society

"Recommended without qualification; and a suitable gift for friends."
Chemical and Engineering News

"There are plenty of useful tips ... in this modestly
priced bijou." *Clinical Chemistry*

"What recommends *THE CHEMIST'S ENGLISH* is
above all its readability ..." *Angewandte Chemie*

"This book is too good to be confined to chemists;
the message is there for all scientists."
Australian Broadcasting Commission

"This book, which may well be one of a kind, is an
utter delight." *Carbohydrate Chemistry*

"The book might well serve as a prescribed text for
PhD students." *Nature*

Date of information:
October 1994

VCH, P.O. Box 10 11 61,
D-69451 Weinheim,
Telefax (0) 62 01 - 60 61 84